T0235793

Noise and Torsional Vibration Analysis of Hybrid Vehicles

Synthesis Lectures on Advances in Automotive Technology

Editor
Amir Khajepour, *University of Waterloo, Canada*

The automotive industry has entered a transformational period that will see an unprecedented evolution in the technological capabilities of vehicles. Significant advances in new manufacturing techniques, low-cost sensors, high processing power, and ubiquitous real-time access to information mean that vehicles are rapidly changing and growing in complexity. These new technologies—including the inevitable evolution toward autonomous vehicles—will ultimately deliver substantial benefits to drivers, passengers, and the environment. Synthesis Lectures on Advances in Automotive Technology Series is intended to introduce such new transformational technologies in the automotive industry to its readers.

Noise and Torsional Vibration Analysis of Hybrid Vehicles
Xiaolin Tang, Yanjun Huang, Hong Wang, and Yechen Qin
2018

Smart Charging and Anti-Idling Systems
Yanjun Huang, Soheil Mohagheghi Fard, Milad Khazraee, Hong Wang, and Amir Khajepour
2018

Design and Avanced Robust Chassis Dynamics Control for X-by-Wire Unmanned Ground Vehicle
Jun Ni, Jibin Hu, and Changle Xiang
2018

Electrification of Heavy-Duty Construction Vehicles
Hong Wang, Yanjun Huang, Amir Khajepour, and Chuan Hu
2017

Vehicle Suspension System Technology and Design
Avesta Goodarzi and Amir Khajepour
2017

Noise and Torsional Vibration Analysis of Hybrid Vehicles
Xiaolin Tang, Yanjun Huang, Hong Wang, and Yechen Qin

ISBN: 978-3-031-00370-7   paperback
ISBN: 978-3-031-01498-7   ebook
ISBN: 978-3-031-00003-4   hardcover

DOI 10.1007/978-3-031-01498-7

A Publication in the Springer series
SYNTHESIS LECTURES ON ADVANCES IN AUTOMOTIVE TECHNOLOGY

Lecture #4
Series Editor: Amir Khajepour, *University of Waterloo, Canada*
Series ISSN
Print 2576-8107   Electronic 2576-8131

Noise and Torsional Vibration Analysis of Hybrid Vehicles

Xiaolin Tang
Chongqing University, China

Yanjun Huang
University of Waterloo, Canada

Hong Wang
University of Waterloo, Canada

Yechen Qin
Beijing Institute of Technology, China

SYNTHESIS LECTURES ON ADVANCES IN AUTOMOTIVE TECHNOLOGY #4

ABSTRACT

Thanks to the potential of reducing fuel consumption and emissions, hybrid electric vehicles (HEVs) have been attracting more and more attention from car manufacturers and researchers. Due to involving two energy sources, i.e., engine and battery, the powertrain in HEVs is a complicated electromechanical coupling system that generates noise and vibration different from that of a traditional vehicle. Accordingly, it is very important to explore the noise and vibration characteristics of HEVs. In this book, a hybrid vehicle with two motors is taken as an example, consisting of a compound planetary gear set (CPGS) as the power-split device, to analyze the noise and vibration characteristics. It is specifically intended for graduates and anyone with an interest in the electrification of full hybrid vehicles.

The book begins with the research background and significance of the HEV. The second chapter presents the structural description and working principal of the target hybrid vehicle. Chapter 3 highlights the noise, vibration, and harshness (NVH) tests and corresponding analysis of the hybrid powertrain. Chapter 4 provides transmission system parameters and meshing stiffness calculation. Chapter 5 discusses the mathematical modeling and analyzes torsional vibration (TV) of HEVs. Finally, modeling of the hybrid powertrain with ADAMS is given in Chapter 6.

KEYWORDS

hybrid, compound planetary gear set, power-split, torsional vibration, noise, transmission efficiency

Contents

Preface

This book explores the noise, vibration, and harshness (NVH) characteristics of a dual-motor hybrid vehicle. It presents the structural description and working principal of a two-motor hybrid vehicle. Then, it analyzes the noise and torsional vibrations (TVs) of the proposed hybrid powertrain. This book is intended for engineers in car companies striving to develop a hybrid vehicle, and graduate and senior undergraduate students in mechanical, automotive, and mechanics engineering. The book is also accessible to anyone interested in learning about the power-split hybrid vehicle.

At first, the background and significance of hybrid electric vehicles (HEVs) are introduced, and then the structural and working principle of the hybrid powertrain is presented. Noise and dynamic modeling with mathematical and ADAMS are also developed and discussed. Moreover, analysis of TVs is carried out and corresponding measures of controlling vibrations are given.

Xiaolin Tang, Yanjun Huang, Hong Wang, and Yechen Qin
December 2018

Acknowledgments

This book would not have been possible without the help of many professionals. We are particularly grateful to Professor Amir Khajepour and Professor Jianwu Zhang for their strong support. We are also thankful to Dr. Haisheng Yu, Dr. Liang Zou for the modeling and experimental verification, and Shanshan Li and Kai Yang for editing and proofreading this book. We are also thankful to Morgan & Claypool Publishers for providing such an opportunity and their consistent encouragement and support.

Xiaolin Tang, Yanjun Huang, Hong Wang, and Yechen Qin
December 2018

Nomenclature

CTD	clutch torsional damper
CPGS	compound planetary gear set
DMF	dual mass flywheel
eCVT	electronic continuously variable transmission
TS	torsional stiffness
MOI	moment of inertia
ω_{S1}	speed of S1, rad/s
ω_{S2}	speed of S2, rad/s
α_b	pressure angle on the index circle, rad
θ	phase angle, rad
θ_{S2}	angular displacement of big sun gear, rad
θ_c	angular displacement of carrier, rad
θ_r	angular displacement of ring gear, rad
θ_e	angular displacement of engine, rad
θ_{aci}	angular displacement of short planet, rad
θ_v	angular displacement of the vehicle, rad
θ_d	angular displacements of the differential, rad
θ_m	angular displacement of the reducer, rad
θ_{rw}	angular displacements of the right wheels, rad
θ_{lw}	angular displacements of the left wheels, rad
θ_{bci}	angular displacement of long planet, rad
ω_{ring}	speed of ring, rad/s
ω_c	speed of carrier, rad/s

ω_1	drive wheel angular velocity, rad/s
z_{ring}	internal tooth number of the ring
z_{S1}	tooth numbers of S1
z_{S2}	tooth numbers of S2
α_1 and α_2	stationary gear ratio
T_{S1}	torque applied on S1, N · m
T_{S2}	torque applied on S2, N · m
$T_{carrier}$	torque applied on carrier, N · m
T_{ring}	torque applied on ring, N · m
T_{E1}	driving torque of E1, N · m
T_{E2}	driving torque of E2, N · m
T_{engine}	driving torque of ICE, N · m
T_L	resistant moment for driving, N · m
J_{S1}	moment of inertia of S1, kg · m^2
J_{S2}	moment of inertia of S2, kg · m^2
$J_{carrier}$	moment of inertia of carrier, kg · m^2
J_{ring}	moment of inertia of ring, kg · m^2
J_e	the moment of inertia of the engine, kg · m^2
J_m	moment of inertia of the reducer, kg · m^2
J_d	moment of inertia of the differential, kg · m^2
J_{lw}	moment of inertia of the left wheel, kg · m^2
J_{rw}	moment of inertia of the right wheel, kg · m^2
I_{f1}	the primary flywheels' rotational inertia, kg · m^2
I_{f2}	the secondary flywheels' rotational inertia, kg · m^2
I_{TCD}	original flywheel's rotational inertia, kg · m^2
α_{S1}	accelerations of S1, m/s^2
α_{S2}	accelerations of S2, m/s^2

$\alpha_{carrier}$	accelerations of carrier, m/s²
α_{ring}	accelerations of ring, m/s²
P_B	battery power, kW
P_{VB}	loss power of battery, kW
P_{VE1}	loss power of E1, kW
P_{VE2}	loss power of E2, kW
P_T^c	friction loss in the transformation mechanism, kW
a_r	angular acceleration of the ring, rad/s²
a_c	angular acceleration of the carrier, rad/s²
δ_{Br}	deformation of the rectangular part
δ_{Bt}	amount of bending deformation of the trapezoidal part
δ_S	amount of deformation caused by shear
δ_G	amount of deformation caused by the inclination of the base part
δ_1	the δ value of meshing tooth
δ_2	the δ value of meshing tooth
δ_{pv}	amount of deformation of the contact surface of the tooth surface
Δt_{oc}	synthetic base error of the driving gear and the driven gear
Δf_p	base section error of the drive gear
Δf_g	base section error of the driven gear
q_{E1}	flexibility for engaging gear pairs
t_c	the impact time, s
C	the velocity of sound, m/s
k_{S2b}	the meshing stiffness between big sun gear and long planet, N/m
k_{ab}	the meshing stiffness between short planet and long planet, N/m
k_{S1a}	the meshing stiffness between small sun gear and short planet, N/m
k_{ar}	the meshing stiffness between ring and short planet, N/m
k_1	the tooth stiffness of the drive gears, N/m

k_2	the tooth stiffness of the driven gears, N/m
k_t	CPG stiffness of the torsional damper, N/m
k_c	CPG stiffness of and the planetary carrier, N/m
k_{rm}	meshing stiffness between the ring and the reducer, N/m
k_{md}	meshing stiffness between the reducer and the differential, N/m
k_{rh}	torsional stiffness of the right half shafts, N/m
k_{lh}	torsional stiffness of the left half shaft, N/m
k_w	torsional stiffness of the wheel, N/m
K	stiffness matrices of 16 order, N/m
k_{pspl}	mesh stiffness between the long planet and sun gear 2, N/m
k_{S1ps}	mesh stiffness between the sun gear 1, and short plane, N/m
k_{rps}	mesh stiffness between the short planet and ring, N/m
k_{CTD}	CTD's torsional stiffness, N/m
k_{DMF}	DMF's torsional stiffness, N/m
M_a	mass of planet a, kg
M_b	mass of planets b, kg
M	mass of 16 order, kg
m_{car}	vehicle mass, kg
r	distance from the point of space to the curvature center of the tooth profile, mm
a	the radius of curvature of the gear meshing impact point, mm
b	tooth width, mm
r_{g1}	basic circle radius of the driving gear, mm
r_{g2}	equivalent base circle radius of a passive gear, mm
r_{short}	radius of the long planet's pitch circle, mm
r_{long}	radius of the long planet's pitch circle, mm
r_{ca}	radius of revolution of the short long planet gears, mm

r_{cb}	radius of revolution of the long planet gears, mm
r_{r2}	base radius of the outer ring, mm
r_w	radius of the wheel, mm
r_{ring}	radius of ring gear's the pitch circle, mm
r_{wheel}	wheel radius, mm
r_{m1} and r_{m2}	base radiuses of reducer gears, mm

CHAPTER 1

Introduction

Since the beginning of the 21st century, resource depletion and environmental pollution have become the two major issues of human survival and development. Resource conservation and environmental protection have become a global consensus [1–8]. Automobiles have changed people's lifestyle and brought development and progress to mankind. However, automobiles also consume a lot of oil resources and cause noise and environmental pollutions. Carbon dioxide emissions from automobile, industrial enterprise, and fossil fuel combustion are the main causes of global warming. For instance, carbon dioxide emissions from vehicles account for more than 60% of all carbon dioxide emissions. In order to achieve sustainable development and reduce carbon dioxide emissions, it is necessary to control the energy consumptions and reduce the dependence on petroleum [9–14]. Accordingly, the universities, carmakers, and research institutes put their emphasis on developing hybrid vehicles with a higher fuel economy and lower emission to replace traditional vehicles [15–21].

Hybrid electric vehicles (HEVs) driven by an internal combustion engine (ICE) and one or more electric motors have been well known for a long time. With the advantages of fuel economy, emissions, and environmental protection, an HEV is taken as one of the most popular traffic tools. In addition, lots of vehicle manufacturers, research departments, and universities worldwide have devoted their efforts to developing HEVs [22–31].

The compound planetary gear set (CPGS) of an HEV is able to combine the power from the ICE, MG1, and MG2, and then provide the power via the ring to achieve power-split. In addition, it can also act as an electronic continuously variable transmission (eCVT) [32–36]. Compared to the traditional Ravingneaux gear set, the proposed system can improve the lever efficiency to reduce the required power of MG2 and save costs. By manipulating the torque and speed of the electric motors, the engine is able to work in the high-efficiency region in the hybrid mode [37–42]. Figure 1.1 shows the schematic of the dual motors HEV.

To identify abnormal noise sources for the HEV with the power-split transmission in different modes, acoustic levels and TVs are measured in a powertrain test bench. The Leuven Measurement & System (LMS) data acquisition instrument and corresponding software are adopted to acquire acoustic and vibration information of the hybrid powertrain. Pressure sensors, acceleration sensors, and photoelectric sensors are also located to acquire the acoustic and vibration information in both the pure and hybrid driving mode. By means of spectral analysis and order tracking, the main noise sources are found and presented.

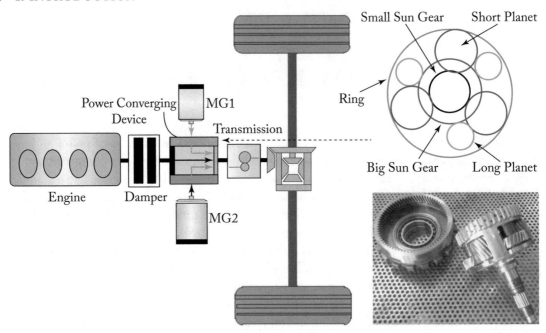

Figure 1.1: Schematic of the dual motors hybrid electric vehicle [20].

The gear meshing of the CPGS is the main noise source in a deep HEV during the pure electric driving mode. Based on the theory of gear meshing noise, meshing noises of gear pairs in the CPGS and hybrid driveline are computed. The meshing noise responses of different pitch error, tooth error, gear speed, and meshing stiffness are analyzed. The results show that the errors of the short planet and the long planet in the CPGS can lead to a higher noise. Therefore, the pitch error and tooth error of the short planet and the long planet should be reduced in producing the compound planetary gear set. This study can provide a theoretical basis for reducing gear meshing noise.

Based on the configuration, dynamic equations of the CPGS are derived by the Lagrangian method. Moreover, the dynamic equations of hybrid driveline including engine, reducer, differential, half shafts, wheels, and vehicle are also derived. Combining those dynamic equations, the mathematical model of the HEV is established. The theoretical predictions of natural frequencies and corresponding vibration modals of the proposed hybrid powertrain under different driving modes are presented.

A torsional model is also built in ADAMS, a commercial software, to acquire the TV characteristics of the studied HEV. The vibration modals and natural frequencies obtained from ADAMS and mathematical model show a remarkable agreement. Therefore, two models are good enough in representing torsional characteristics of the hybrid powertrain. Moreover, the main factors affecting the TV under different excitations are studied by the forced vibration analysis. Optimal parameters are also provided, where the amplitude of TV of the powertrain is minimized.

A dual mass flywheel torsional damper is introduced to the powertrain to analyze TV characteristics. Comparing to simulation results, it is found that the dual mass flywheel torsional damper can effectively reduce TVs of the hybrid powertrain. In addition, the different rotational inertia of the first and second flywheel, torsional stiffness (TS), and damping are also considered in this research and at last, their optimal values are obtained and presented.

This study can definitely be used as a valuable reference when developing an HEV.

CHAPTER 2

Structural Description and Work Principle of Full Hybrid Vehicles

Figure 2.1 shows the configuration of the driveline of a power-split HEV with the CPGS, consisting of an ICE, CPGS, torsional damper (D), reducer (R), electric motors ($E1$ and $E2$), differential ($Diff$), drive shafts, and wheels. In addition, two brakes $B1$ and $B2$ are used to lock the carrier and small sun gear $S1$, respectively. More specifically, the front view of the CPGS of the four-shaft transmission with $E1$ and $E2$ is indicated in Fig. 2.2 as well [49].

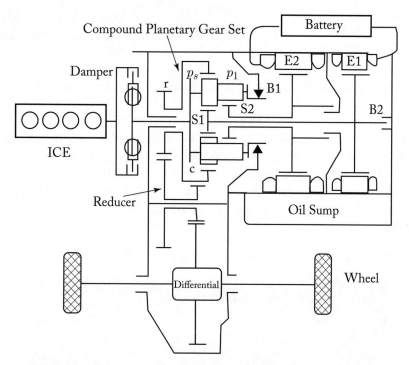

Figure 2.1: Configuration of the power-split hybrid powertrain.

As shown in Fig. 2.2, the CPGS acts as a power-split component in the hybrid powertrain. It combines the power from the engine, $E1$, and $E2$, and drives wheels via the ring, reducer, and differential. In addition, it also performs as an eCVT. The CPGS includes a ring (r), a carrier (c), a small sun gear 1 $(S1)$, a big sun gear 2 $(S2)$, three short planets (p_s), and three long planets (p_l). All planets are attached to (c) and the short planet and the corresponding long planet engage with each other. $S1$ and $S2$ connects $E1$ and $E2$, respectively, and c is attached to the engine via the torsional damper. Additionally, to realize the optimal control and have the engine operate in high-efficiency regions, the powertrain utilizes two brakes to stop $S1$ and the carrier, respectively.

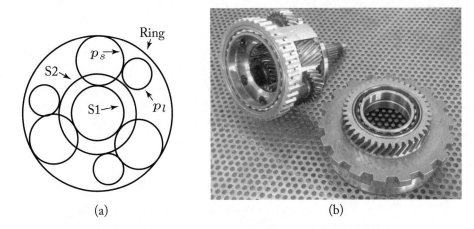

(a)   (b)

Figure 2.2: Structure of the CPGS: (a) axial view and (b) geometry of the CPGS.

For the sake of analysis, the CPGS is split into two rows, where the first is composed of $S1$, short planets, and the ring; while the second includes $S2$, short and long planets, and the ring. The speed of the CPGS can be obtained with the tabular method as follows: first, bind and rotate the CPGS with ω_c, the speed of carrier; second, lock the carrier, release others, and then let the ring (or sun gear) rotate with $\omega_{ring} - \omega_c$. The speed of the CPGS in the third row of Tables 2.1 and 2.2 can be obtained from the speed correlation between the first and second row of the CPGS [50].

Table 2.1: The speed correlation of the first gear group in the CPGS

	Carrier	Ring	Sun Gear 1
1	ω_c	ω_c	ω_c
2	0	$\omega_{ring} - \omega_c$	$-(\omega_{ring} - \omega_c) \cdot \alpha_1$
3	ω_c	ω_{ring}	$\omega_c(1 + \alpha_1) - \omega_{ring} \cdot \alpha_1 = \omega_{s1}$

Table 2.2: The speed correlation of the second gear group in the CPGS

	Carrier	Ring	Sun Gear2
1	ω_c	ω_c	ω_c
2	0	$\omega_{ring} - \omega_c$	$(\omega_{ring} - \omega_c) \bullet \alpha_2$
3	ω_c	ω_{ring}	$\omega_c(1 - \alpha_2) + \omega_{ring} \bullet \alpha_2 = \omega_{s2}$

As shown in Tables 2.1 and 2.2, reformulate the speeds of the CPGS as follows:

$$\omega_{S1} + \alpha_1 \cdot \omega_{ring} = (1 + \alpha_1)\omega_c \qquad \alpha_1 = \frac{z_{ring}}{z_{S1}} \tag{2.1}$$

$$\omega_{S2} - \alpha_2 \cdot \omega_{ring} = (1 - \alpha_2)\omega_c \qquad \alpha_2 = \frac{z_{ring}}{z_{S2}}, \tag{2.2}$$

where ω_{S1} and ω_{S2} denote the speed of $S1$ and $S2$, respectively; α_1 and α_2 refer to the stationary gear ratios; ω_{ring} and ω_c are the speed of ring and carrier; z_{ring} means the internal tooth number of the ring; z_{S1} and z_{S2} are tooth numbers of $S1$ and $S2$, respectively. Based on the structure of the CPGS, the speed of $E1$, $E2$, engine, and ring can be obtained:

$$\omega_{E1} = (1 + \alpha_1)\omega_c - \alpha_1 \cdot \omega_{ring} \tag{2.3}$$

$$\omega_{E2} = (1 + \alpha_2)\omega_c + \alpha_2 \cdot \omega_{ring} \tag{2.4}$$

$$\omega_{engine} = \frac{\alpha_1 \omega_{S2} + \alpha_2 \omega_{S1}}{\alpha_1 + \alpha_2} \tag{2.5}$$

$$\omega_{ring} = \frac{(\alpha_2 - 1)\omega_{S1} - (\alpha_1 + 1)\omega_{S2}}{\alpha_1 + \alpha_2}. \tag{2.6}$$

The equilibrium equation of CPGS power is rewritten as:

$$T_c\omega_c + T_{ring}\omega_{ring} + T_{S1}\omega_{S1} + T_{S2}\omega_{S2} = 0. \tag{2.7}$$

Equation (2.5) shows the speed of engine does not rely on the vehicle speed. By manipulating the torque and speed of $E1$ and $E2$, the engine is able to operate in high-efficiency regions during the hybrid mode. In addition, continuous various speeds and a lower fuel consumption are reached while the driving force is met. The working principle can be described clearly by lever principle to draw a parallel. As depicted in Fig. 2.3, once the engine speed is set, the ring speed is varying with speeds of electric motors. In other words, the ring speed can be continuously varied via manipulating the motors.

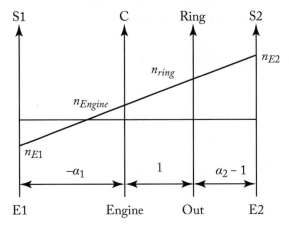

Figure 2.3: The lever principle for the CPGS.

The following equilibrium equations can be obtained:

$$T_{S1} - T_{E1} + J_{S1} \cdot \alpha_{S1} = 0 \tag{2.8}$$

$$T_{S2} - T_{E2} + J_{S2} \cdot \alpha_{S2} = 0 \tag{2.9}$$

$$T_{engine} - T_{carrier} - J_{carrier} \cdot \alpha_{carrier} = 0 \tag{2.10}$$

$$T_{ring} + T_L + J_{ring} \cdot \alpha_{ring} = 0, \tag{2.11}$$

where $T_{S1}, T_{S2}, T_{carrier}, T_{ring}$ denote the torque applied on $S1$, $S2$, carrier, and ring, respectively. $T_{E1}, T_{E2}, T_{engine},$ and T_L denote the driving torque of $E1$, $E2$, ICE, and loading torque, respectively. $J_{S1}, J_{S2}, J_{carrier},$ and J_{ring} refer to the moment of inertia of $S1$, $S2$, carrier, and ring, respectively. $\alpha_{S1}, \alpha_{S2}, \alpha_{carrier},$ and α_{ring} indicate the accelerations of $S1$, $S2$, carrier, and ring, respectively.

The equilibrium equations of the torques of the whole system are:

$$T_{carrier} + T_{S1} + T_{S2} + T_{ring} = 0 \tag{2.12}$$

$$T_{ring} + T_{S1} \cdot \alpha_1 + T_{S2} \cdot \alpha_2 = 0 \tag{2.13}$$

$$T_{carrier}\omega_{carrier} + T_{ring}\omega_{ring} + T_{S1}\omega_{S1} + T_{S2}\omega_{S2} = 0. \tag{2.14}$$

Power conservation in the whole system can be represented by:

$$P_B + P_V + P_{VE1} + P_{VE2} = \frac{\pi}{30} \cdot (T_{E1} \cdot \omega_{S1} + T_{E2} \cdot \omega_{S2}),\tag{2.15}$$

where P_B means the battery power. P_V, P_{VE1}, and P_{VE2} refer to the loss power of battery and $E1$ and $E2$, respectively.

The e-CVT transmits the power by the CPGS instead of hydraulic devices and thus the efficiency can reach as much as 95%. Its efficiency under some typical operating conditions is presented and it is an important criterion when evaluating the properties of a transmission system [43–48]. There are two key characteristics: first, the efficiencies of planetary gear trains depend on their structure type; and second, the efficiency of planetary gear trains of the same type varies with the transmission ratio and with changes in inputs and outputs. The efficiency of a compound planetary gear transmission in a deep HEV is therefore complex to calculate.

2.1 THE EFFICIENCY OF THE TRANSMISSION SYSTEM

An efficiency analysis of the power converging transmission under a range of typical working conditions as follows.

2.1.1 HYBRID DRIVING MODE

In this mode, the vehicle is powered by the engine and two motors via the ring. A lever diagram of this mode is represented by Fig. 2.4.

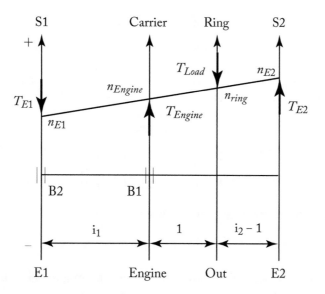

Figure 2.4: Torque balance diagram.

The efficiency in this mode can be analyzed under a range of different conditions.

1. When $\omega_{S2} > \omega_r > \omega_c > \omega_{S1} > 0$.

The input power of the CPGS is:

$$P_c = T_c \cdot \omega_c > 0 \tag{2.16}$$

$$P_{S1} = T_{S1} \cdot \omega_{S1} > 0 \tag{2.17}$$

$$P_{S2} = T_{S2} \cdot \omega_{S2} > 0. \tag{2.18}$$

The output power can be defined as

$$P_r = T_r \cdot \omega_r < 0. \tag{2.19}$$

Equations (2.16)–(2.19) for $\omega_{S2} > \omega_r > \omega_c > \omega_{S1} > 0$ can be written as

$$T_c > 0, \ T_{S1} > 0, \ T_{S2} > 0, \ T_r < 0. \tag{2.20}$$

The efficiency of the CPGS can be written as

$$
\begin{aligned}
\eta_{(S1,S2,c)r} &= \frac{|P_r|}{P_{S1} + P_{S2} + P_c} \\
&= \frac{|T_r \cdot \omega_r|}{T_{S1} \cdot \omega_{S1} + T_{S2} \cdot \omega_{S2} + T_c \cdot \omega_c}.
\end{aligned}
\tag{2.21}
$$

Given the speed of the CPGS, the following equations are obtained:

$$\omega_{S1}^c = \omega_{S1} - \omega_c < 0 \tag{2.22}$$

$$\omega_{S2}^c = \omega_{S2} - \omega_c > 0 \tag{2.23}$$

$$\omega_r^c = \omega_r - \omega_c > 0. \tag{2.24}$$

The meshing power of the sun gear 1 can be written as

$$P_{S1}^c = T_{S1}\,(\omega_{S1} - \omega_c) < 0. \tag{2.25}$$

The meshing power of the sun gear 2 can be written as

$$P_{S2}^c = T_{S2}\,(\omega_{S2} - \omega_c) > 0. \tag{2.26}$$

The meshing power of the ring gear can be written as

$$P_r^c = T_r\,(\omega_r - \omega_c) < 0. \tag{2.27}$$

From Eqs. (2.25)–(2.27), under these conditions, the sun gear 2 is the input power component while sun gear 1 and the ring gear are output power components. Meshing power flows from sun gear 2 to the ring and sun gear 1.

The power balance equation for the transformation mechanism of the CPGS can be expressed as

$$T_{S1} \cdot \omega_{S1}^c + T_{S2} \cdot \omega_{S2}^c \cdot \eta_{S2(S1,r)}^c + T_r \cdot \omega_r^c = 0, \tag{2.28}$$

where $\eta_{S2(S1,r)}^c$ is the efficiency of the transformation mechanism of the CPGS, defined as:

$$\eta_{S2(S1,r)}^c = \frac{-P_{S1}^c - P_r^c}{P_{S2}^c}. \tag{2.29}$$

The gear ratio can be written as

$$i_{S1,r}^c = \frac{\omega_{S1}^c}{\omega_r^c} = -\alpha_1 \tag{2.30}$$

$$i_{S2,r}^c = \frac{\omega_{S2}^c}{\omega_r^c} = \alpha_2. \tag{2.31}$$

Inserting Eqs. (2.30) and (2.31) into Eq. (2.28) comes to

$$T_{S1} \cdot (-\alpha_1) + T_{S2} \cdot \alpha_2 \cdot \eta_{S2(S1,r)}^c + T_r = 0. \tag{2.32}$$

The moment equilibrium equation for the CPGS is

$$T_{S1} + T_{S2} + T_c + T_r = 0. \tag{2.33}$$

Combining Eqs. (2.31), (2.32), and (2.33), we can derive

$$\eta = \frac{\left| \left(\alpha_1 \cdot T_{S1} - \alpha_2 \cdot T_{S2} \cdot \eta_{S2(S1,r)}^c \right) \omega_r \right|}{T_{S1} \cdot \omega_{S1} + T_{S2} \cdot \omega_{S2} + \left(-T_{S1} - T_{S2} - \alpha_1 \cdot T_{S1} + \alpha_2 \cdot T_{S2} \cdot \eta_{S2(S1,r)}^c \right) \omega_c}$$
$$= \frac{\left| \left(\alpha_1 \cdot T_{S1} - \alpha_2 \cdot T_{S2} \cdot \eta_{S2(S1,r)}^c \right) \omega_r \right|}{T_{S1} \left(\omega_{S1} - \omega_c - \alpha_1 \cdot \omega_c \right) + T_{S2} \left(\omega_{S2} - \omega_c + \alpha_2 \cdot \eta_{S2(S1,r)}^c \cdot \omega_c \right)}, \tag{2.34}$$

where $T_{S1} = T_{E1} - J_{S1} \cdot a_{S1}$, $T_{S2} = T_{E2} - J_{S2} \cdot a_{S2}$, T_{E1} and T_{E2} are the driving torque of $E1$ and $E2$, respectively. J_{S1} and J_{S2} refer to the moment of inertia of $S1$ and $S2$, respectively. a_{S1} and a_{S2} denote the acceleration of $S1$ and $S2$, respectively.

2. When $\omega_{S2} > \omega_r > \omega_c > 0 > \omega_{S1}$.

   According to Eq. (2.9), the meshing power of the sun gear 1 is:

$$P_{S1}^c = T_{S1} \left(\omega_{S1} - \omega_c \right) > 0. \tag{2.35}$$

   The meshing power of the sun gear 2 can be written as

$$P_{S2}^c = T_{S2} \left(\omega_{S2} - \omega_c \right) > 0. \tag{2.36}$$

The meshing power of the ring gear can be written as

$$P_r^c = T_r \left(\omega_r - \omega_c \right) < 0. \tag{2.37}$$

According to Eqs. (2.35) and (2.36), under these conditions, $S2$ and $S1$ are the input power component, while the ring gear is the output power component.

The efficiency of the transformation mechanism of the CPGS can be defined as

$$\eta_{(S1,S2)r}^c = \frac{|P_r^c|}{P_{S1}^c + P_{S2}^c}. \tag{2.38}$$

The power balance equation can be expressed as

$$T_{S1} \cdot \omega_{S1}^c \cdot \eta_{(S1,S2)r}^r + T_{S2} \cdot \omega_{S2}^c \cdot \eta_{(S1,S2)r}^c + T_r \cdot \omega_r^c = 0. \tag{2.39}$$

That combining Eqs. (2.30), (2.31), and (2.38) gives

$$T_{S1} \cdot (-\alpha_1) \cdot \eta_{(S1,S2)r}^c + T_{S2} \cdot \alpha_2 \cdot \eta_{(S1,S2)r}^c + T_r = 0. \tag{2.40}$$

Combining Eqs. (2.10), (2.23), and (2.40), the efficiency is given by

$$\begin{aligned}
\eta &= \frac{|T_r \cdot \omega_r|}{T_{S1} \cdot \omega_{S1} + T_{S2} \cdot \omega_{S2} + T_c \cdot \omega_c} \\
&= \frac{\left| (\alpha_1 \cdot T_{S1} - T_{S2} \cdot \alpha_2) \, \eta_{(S1,S2)r}^r \cdot \omega_r \right|}{T_{S1} \left(\omega_{S1} - \omega_c - \alpha_1 \cdot \eta_{(S1,S2)r}^r \cdot \omega_c \right)} \\
&\quad + T_{S2} \left(\omega_{S2} - \omega_c + \alpha_2 \cdot \eta_{(S1,S2)r}^r \cdot \omega_c \right).
\end{aligned} \tag{2.41}$$

2.1.2   PURE ELECTRIC DRIVING MODE

During this mode, the carrier and ICE are locked and the vehicle is powered by two electric motors. Therefore, the power inputs from electric motors and outputs from the ring. The lever diagram is indicated in Fig. 2.5.

The efficiency of the CPGS under the pure electric mode can be defined as

$$\begin{aligned}
\eta &= \frac{-P_r}{P_{S1} + P_{S2}} \\
&= \frac{|T_r \cdot \omega_r|}{T_{S1} \cdot \omega_{S1} + T_{S2} \cdot \omega_{S2}}.
\end{aligned} \tag{2.42}$$

The speed of the CPGS can be written as $\omega_{S2} > \omega_r > 0 > \omega_{S1}$.

The torque loading on the ring can be expressed as

$$T_r = T_L + J_r \cdot a_r, \tag{2.43}$$

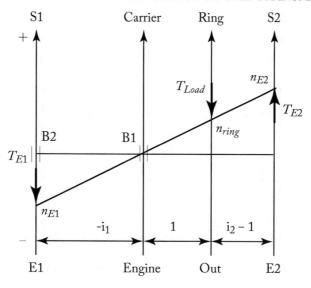

Figure 2.5: Torque balance diagram.

where T_L is the resistant moment for driving, J_r denotes the inertia of the ring, and a_r is the angular acceleration of the ring.

The torque loading on $S1$ and $S2$ is:

$$T_{S1} = T_{E1} - J_{S1} \cdot a_{S1} \tag{2.44}$$

$$T_{S2} = T_{E2} - J_{S2} \cdot a_{S2}. \tag{2.45}$$

Submitting Eqs. (2.43), (2.44), and (2.45) into (2.42), and the efficiency of the CPGS under pure electric mode can be expressed as

$$\eta = \frac{|(T_L + J_r \cdot a_r)\omega_r|}{(T_{E1} - J_{S1} \cdot a_{S1})\omega_{S1}} + (T_{E2} - J_{S2} \cdot a_{S2})\omega_{S2}. \tag{2.46}$$

2.1.3   OVER SPEED DRIVING MODE

In this mode, the vehicle is powered by the ICE and $E2$. The lever diagram is presented in Fig. 2.6.

The input power of the CPGS can be written as

$$P_c = T_c \cdot \omega_c > 0 \tag{2.47}$$

$$P_{S2} = T_{S2} \cdot \omega_{S2} > 0. \tag{2.48}$$

The output power for the CPGS can be defined as

$$P_r = T_r \cdot \omega_r < 0. \tag{2.49}$$

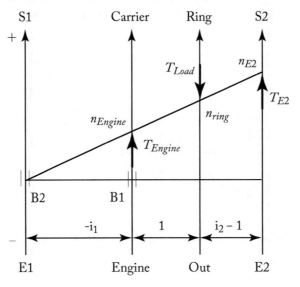

Figure 2.6: Torque balance diagram.

The speed correlation for the CPGS is

$$\omega_{S2} > \omega_r > \omega_c > \omega_{S1} = 0. \tag{2.50}$$

Sun gear 2 in the transformation mechanism of the CPGS can be defined as

$$P_{S2}^c = T_{S2} (\omega_{S2} - \omega_c) > 0. \tag{2.51}$$

The meshing power of the ring gear can be written as

$$P_r^c = T_r (\omega_r - \omega_c) < 0. \tag{2.52}$$

According to Eqs. (2.51) and (2.52), in this mode, $S2$ is the input power component and the ring is output power component. The meshing power of sun gear 2 is divided into two types.

1. When $P_{S1}^c = (\omega_{S1} - \omega_c) T_c > 0$.

   The efficiency of the transformation mechanism of the CPGS can be expressed as

$$\eta_{(S1,S2)r}^c = \frac{|P_r^c|}{|P_r^c| + P_T^c}$$

$$= \frac{1}{1 + \dfrac{P_T^c}{|P_r^c|}}, \tag{2.53}$$

where P_T^c is the friction loss in the transformation mechanism.

The power ratio of the transformation mechanism and ring is

$$\frac{P_r^c}{P_r} = \frac{T_r(\omega_r - \omega_c)}{T_r \cdot \omega_r} = 1 - \frac{\omega_c}{\omega_r} = 1 - \frac{i_{S1,r}^c}{i_{S1,r}^c - 1} = \frac{1}{1 + \alpha_1}. \tag{2.54}$$

Therefore,

$$P_r^c = \frac{1}{1 + \alpha_1} P_r. \tag{2.55}$$

Because the loss of power in the transformation mechanism equals to actual loss power, Eqs. (2.53) and (2.55) combined to give

$$P_T^c = \frac{1 - \eta_{(S1,S2)r}^c}{\eta_{(S1,S2)r}^c} \cdot |P_r^c| = \frac{1 - \eta_{(S1,S2)r}^c}{\eta_{(S1,S2)r}^c} \cdot \left| \frac{1}{1 + \alpha_1} P_r \right| = P_T. \tag{2.56}$$

The efficiency of the CPGS in the over speed driving mode can be defined as

$$\eta = \frac{1}{1 + \dfrac{P_T}{|P_r|}}. \tag{2.57}$$

Substituting Eq. (2.36) into (2.47), the efficiency can be restated as

$$\eta = \frac{1 - \eta_{(S1,S2)r}^c}{\eta_{(S1,S2)r}^c \cdot (1 + \alpha_1)}. \tag{2.58}$$

2. When $P_{S1}^c = (\omega_{S1} - \omega_c) T_c < 0$.

Under these conditions, the efficiency of the transformation mechanism of the CPGS can be expressed as

$$\eta_{S2(S1,r)}^c = 1 - \frac{P_T^c}{P_{S2}^c}. \tag{2.59}$$

The power ratio between the transformation mechanism and reality for sun gear 2 can be written as

$$\frac{P_{S2}^c}{P_{S2}} = \frac{T_{S2}(\omega_{S2} - \omega_c)}{T_{S2} \cdot \omega_{S2}} = 1 - \frac{\omega_c}{\omega_{S2}}$$
$$= 1 - i_{c,S2}^r = \frac{i_{S2,r}^c}{i_{S2,r}^c - 1} = \frac{\alpha_2}{\alpha_2 - 1}. \tag{2.60}$$

Therefore,

$$P_{S2}^c = \frac{\alpha_2}{\alpha_2 - 1} P_{S2}. \tag{2.61}$$

As the loss power in the transformation mechanism equals to actual loss power, Eqs. (2.59) and (2.61) can be combined to give

$$P_T^c = \left(1 - \eta_{S2(S1,r)}^c\right) \frac{\alpha_2}{\alpha_2 - 1} P_{S2} = P_T.$$ (2.62)

The efficiency of the CPGS in the over speed driving mode can be expressed as

$$
\begin{aligned}
\eta &= \frac{1}{1 + \dfrac{P_T}{|P_r|}} \\
&= \frac{|P_r| \cdot (\alpha_2 - 1)}{|P_r| \cdot (\alpha_2 - 1) + \left(1 - \eta_{S2(S1,r)}^c\right) \cdot \alpha_2 \cdot P_{S2}} \\
&= \frac{|(T_L + J_r \cdot a_r) \cdot \omega_r| \cdot (\alpha_2 - 1)}{|(T_L + J_r \cdot a_r) \cdot \omega_r| \cdot (\alpha_2 - 1) + \left(1 - \eta_{S2(S1,r)}^c\right) \cdot \alpha_2 \cdot (T_{E2} - J_{S2} \cdot a_{S2}) \cdot \omega_{S2}}.
\end{aligned}
$$ (2.63)

2.1.4   STANDSTILL CHARGING MODE

In this mode, when the vehicle stops, the ICE charges the battery. The lever diagram is indicated in Fig. 2.7.

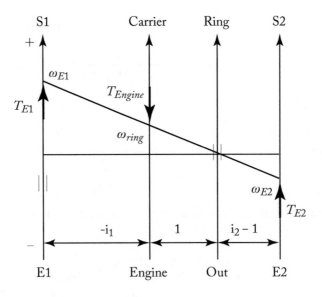

Figure 2.7: Torque balance diagram.

During this mode, the speed correlation for the planetary gear components is

$$\omega_{S1} > \omega_c > 0 > \omega_{S2}.$$ (2.64)

The input power of the CPGS is

$$P_c = T_c \cdot \omega_c > 0. \tag{2.65}$$

The output power for the CPGS can be defined as

$$P_{S1} = T_{S1} \cdot \omega_{S1} < 0 \tag{2.66}$$

$$P_{S2} = T_{S2} \cdot \omega_{S2} < 0. \tag{2.67}$$

The efficiency of the CPGS in the standstill mode can be expressed as

$$\eta = \frac{-P_{S1} - P_{S2}}{P_c}$$
$$= \frac{|T_{S1} \cdot \omega_{S1}| + |T_{S2} \cdot \omega_{S2}|}{T_c \cdot \omega_c}. \tag{2.68}$$

The torque loaded on the carrier can be defined as

$$T_{engine} - J_c \cdot a_c = T_c, \tag{2.69}$$

where T_{engine} denotes the driving torque of the engine; J_c refers to the inertia of the carrier and a_c is the angular acceleration of the carrier.

Combining Eqs. (2.44), (2.45), and (2.68), the efficiency of the CPGS in the standstill mode can be written as

$$\eta = \frac{|(T_{E1} - J_{S1} \cdot a_{S1}) \cdot \omega_{S1}| + |(T_{E2} - J_{S2} \cdot a_{S2}) \cdot \omega_{S2}|}{(T_{engine} - J_c \cdot a_c) \cdot \omega_c}. \tag{2.70}$$

2.2   CONCLUSION

The dynamic equations for the transmission and the speed relationships of a power-split device of the series-parallel type have been derived, and the efficiency of the power-split HEV has been calculated under a range of working conditions: hybrid driving mode, pure electric driving mode, over speed driving mode, and standstill charging mode. This section can provide a valuable reference to guide the control design for power and efficiency of HEVs.

CHAPTER 3

NVH Testing and Analysis of Hybrid Powertrains

The noise and vibration are attracting more and more attention, and it has become one of the most important technical indicators of vehicles. Moreover, many countries establish strict rules to control the vibration and noise of automobiles, and many consumers take the NVH performance as one of the important considerations when choosing a car [49–55]. Therefore, it is necessary to conduct the vibration and noise tests of hybrid vehicles, to identify their main noise sources and make an objective evaluation of their vibration and noise conditions.

Compared to traditional cars, the full HEV is a complex electromechanical coupling system, which produces vibration and noise different from traditional vehicles. In order to solve the serious noise and vibration problem of the hybrid powertrain, experiments are carried out and test results are analyzed.

3.1 PREPARE OF EXPERIMENTS

This experiment is based on the above-mentioned compound planetary power-split hybrid system developed by Greely Automobile Co., Ltd, which is a power-coupled transmission consisting of an engine and two motors. It uses a two-row planetary gear set, two independent high-efficiency DC motor, and two lock-up brakes to make the system effective. The vibration and noise of this system are within the acceptable range of the human body during the hybrid mode. However, there is a serious NVH issue under the pure electric mode, which directly affects the ride comfort. To solve these problems, it is necessary to carry out tests to figure out the main noise source and make an objective evaluation of the vibration and noise conditions of the powertrain.

3.1.1 TEST ENVIRONMENT

The HEV powertrain experimental system is provided in Fig. 3.1, consisting of a 1.6 L ICE, the CPGS, $E1$ and $E2$, Nickel hydrogen battery package, air cooling fan, gasoline supply, analogical loading AC motors, electric units, and sensors. The experimental system is attached to loading motors to simulate a vehicle. The engine's maximum power is 89 kW with a nominal speed of 5500 rpm and a maximum torque of 150 Nm. In addition, the battery utilizes 192 cells with a nominal voltage of 1.5 V. The battery connects to $E1$ and $E2$, permanent magnetic synchronous

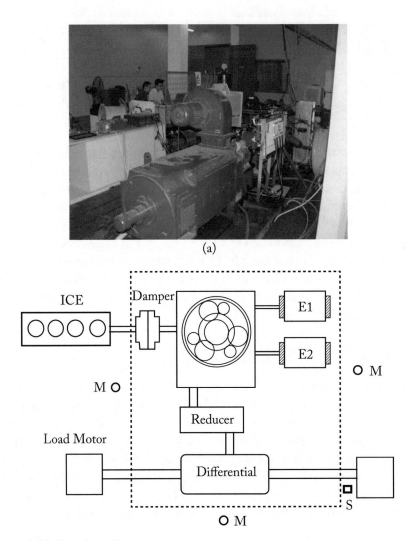

M: Microphone Location
S: Speed Sensor Location

(b)

Figure 3.1: Test rig and measurement position [36].

machines, through a water-cooled power control unit with inverters. The specifications of motor $E1$ and $E2$ are maximum power 42 kW and 57 kW, top speed of 10500 rpm and 8500 rpm, and a max torque of 79 Nm and 190 Nm, respectively. Both motors can switch to generators to charge the battery. The structure of the experimental system with its control system is presented in Fig. 3.2.

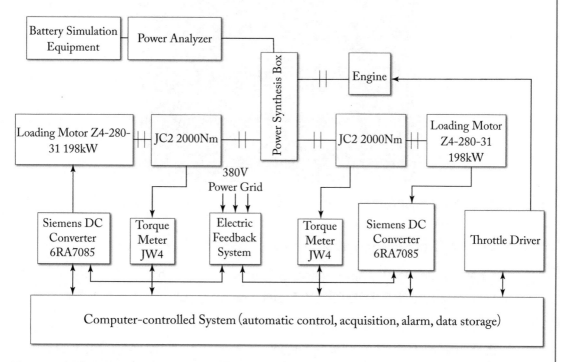

Figure 3.2: Bench test and control system for the hybrid powertrain.

3.1.2 TEST EQUIPMENT

The test instruments and equipment include LMS data instrument and its related signal processing software (LMS SCADAS III), microphones, three directional acceleration sensors, one directional acceleration sensor, and a photoelectric sensor.

1. LMS data acquisition instrument and its post-processing software.

   The noise and vibration tests were conducted by the 24-channel LMS SCADAS III data system. LMS is a leader in NVH testing, with strengths in vibration and noise data acquisition and signal processing, and high-end acoustic testing technology. The versatility, reliability, and advancement of LMS's testing and analysis systems receive the recognition from many automakers and universities. Figure 3.3 shows the LMS test equipment.

Figure 3.3: LMS test equipment.

Modules of LMS hardware system includes:

(a) LMS SCADAS III Data acquisition front end;

(b) 24-channel/ICP/TEDS/BRIDGE input modules; and

(c) 2-channel speed input modules.

Each channel of the LMS has a sampling frequency of 102.4 kHz, 24-bit DSP technology, a signal-to-noise ratio of 105 dB, and a data rate of 2.2 M samples/sec. The LMS system supports all types of sensors related to NVH measurements, including PCB company's ICP type accelerometer, ICP type microphone, handheld accelerometer calibrator, etc.

The LMS software system integrates seamlessly with the above hardware systems, which including the following software modules:

(a) order tracking module;

(b) data post processing module; and

(c) spectral testing module.

Using these hardware and software systems, we can carry out the time domain analysis of noise and vibration, order tracking, video conversion, octave analysis, spectrum analysis, etc.

2. Microphone.

The microphone is used to test the sound pressure of the system, which is transmitted to the LMS data acquisition instrument through the data line, and the data acquisition instrument obtains the system noise spectrum map through post-processing analysis. Figure 3.4 shows the Microphone.

Figure 3.4: Microphone.

3. Acceleration sensor.

Three-direction acceleration and one-direction acceleration sensors are used in this test. The three-direction acceleration sensor can test the acceleration signals in the three directions of x, y, and z. The one-direction acceleration sensor can only test the acceleration signal in one direction. The accelerometer is connected to the LMS data acquisition device via a data line to transmit the signal to the LMS for post-processing. The acceleration sensor used in the experiment is shown in Fig. 3.5.

4. Photoelectric sensor.

In this experiment, the photoelectric speed sensor is adopted to track the rotation speed of the load motor. The signal measured by the photoelectric speed sensor and the sound pressure signal obtained by the microphone test can be used to get the noise spectrum results after LMS processing.

3.1.3 MEASURING POINT ARRANGEMENT

To effectively diagnose the noise source of the hybrid system in pure electric and hybrid mode, it is necessary to carry out the bench test under various typical working conditions. By arranging

Figure 3.5: Acceleration sensor.

the sound pressure sensor next to the hybrid system and arranging the acceleration sensor on the system, the noise data and acceleration data will be obtained, and the noise source of the hybrid system will be verified through spectrum analysis and order tracking, which provides the basis for vibration and noise reduction of the hybrid vehicle.

In the experiment, the measuring points of the sound pressure sensor, the acceleration sensor, and the photoelectric speed sensor are arranged as follows.

1. The arrangement of a sound pressure sensor (Microphone).

   During the test, three sound pressure sensors were used, arranged in the front, right, and bottom of the power-split device to collect the noise signals of the hybrid system under different working conditions. The measurement points of the three-sound pressure sensor are shown in Table 3.1, and the position of the microphone measurement points is depicted in Fig. 3.6.

2. The arrangement of the acceleration sensor.

   To obtain the vibrational acceleration signal of the hybrid synthesis box while testing the noise, eight acceleration sensors were used in this experiment, among which there are four three-direction and one-direction acceleration sensors. The arrangement of the accelerometer measuring points is shown in Table 3.2. The position of some acceleration sensors is shown in Fig. 3.7.

3. The arrangement of the photoelectric speed sensor.

Table 3.1: Locations of sound pressure sensors

Sensor	Measure Point	Measuring Point Position
Microphone 1	MIC_Front	Located in the middle of the front end cover of the power synthesis box, the distance from the box is about 4.8 cm.
Microphone 2	MIC_Right	Located near the center of the right end cover of the power synthesis box, it is 2.6 cm away from the box.
Microphone 3	MIC_Down	Located near the center of the bottom end of the synthetic box, the distance from the box is about 7.7 cm.

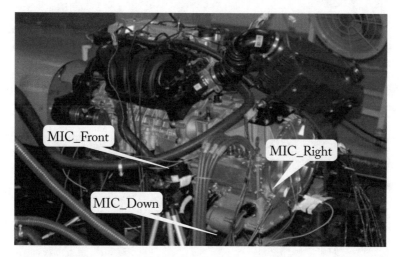

Figure 3.6: Locations of the microphones.

A photoelectric speed sensor was also installed in the experiment to track the order in the noise test. The photoelectric speed sensor is located under the connecting shaft on the left side of the load motor. The specific position is shown in Fig. 3.8 [49].

3.2 DESCRIPTION OF THE TEST PROGRAM AND CONDITIONS

For a full hybrid system, serious noise issues appear under pure electric working conditions. Five typical working conditions tests have been carried out, including four pure electric working modes and one hybrid working mode. Table 3.3 gives a detailed description of the five test cases. Under condition 1 and 2, the rotational speed of the load motor is constant; while in condition 3 and 4, the rotational speed of the load motor is changed. Conditions 2 and 3 are separately driven

Table 3.2: Locations of acceleration sensors

Sensor	Measure Point	Measuring Point Position
Three-direction 1	MEEBS_right	Near the center of the right side cover of the power synthesis box
Three-direction 2	MEEBS_right_up	The upper part of the right side of the power synthesis box.
Three-direction 3	MEEBS_front_up	The front end of the power synthesis box.
Three-direction 4	ENGINE_up	The top left side of the engine.
One-direction 1	SUSP_LF	On the engine, the left end is suspended near the front.
One-direction 2	SUSP_RF	On the engine, the right end is suspended near the front.
One-direction 3	SUSP_RR	On the engine, the right end is hung near the rear.
One-direction 4	SUSP_LR	On the engine, the left end is suspended near the rear.

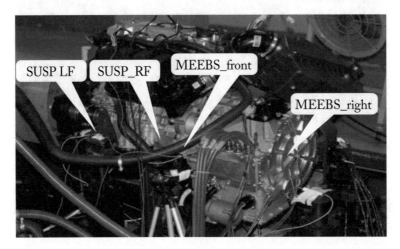

Figure 3.7: Locations of some acceleration sensors.

by the small motor $E1$, and condition 1 and 4 are separately driven by the large motor $E2$. In addition, the condition 5 is a hybrid operating condition in which the hybrid system is started by the motor, and the engine and the motor jointly drive the hybrid system when the load motor speed reaches 350 rpm [36].

Figure 3.8: Location of photoelectric sensor.

Table 3.3: Details of testing conditions

Work Condition	Driving Mode	Description
1	Pure electric mode	Brake B1 is locked, the load motor speed is 250 rpm, E2 drives only and the E2 applied torque is 0–90 Nm.
2	Pure electric mode	Brake B1 is locked, the load motor speed is 350 rpm, E1 drives only and the E1 applied torque is 0–30 Nm.
3	Pure electric mode	Brake B1 is locked, the load motor speed is from 0–350 rpm, E1 drives only and the E1 applied torque is -20 Nm.
4	Pure electric mode	Brake B1 is locked, the load motor speed is from 0–500 rpm, E2 drives only and the E2 applied torque is 60 Nm.
5	Hybrid driving mode	The pedal opening is 35%, the load motor speed is from 150–750 rpm.

3.3 ANALYSIS OF THE EXPERIMENTAL RESULTS

This section lists the vibration and noise test results under five operating conditions. It mainly analyzes the noise sources based on the noise results from the microphone test. It also studies the frequency spectrum and order tracking of the noise and calculates the gear mesh frequency. The main noise source of the power synthesis box is figured out. Compared with the results of the natural frequency of the hybrid power transmission in the following discussion, it can be seen that the noise caused by TV is also an important noise source.

3.3.1 TEST RESULTS OF WORKING CONDITION 1

Condition 1 represents the pure electric mode, where brake B1 is locked and the load motor speed is 250 rpm. Only motor $E2$ works and its torque is 0–90 Nm. Figure 3.9 presents results of the microphone 1 located in front of the power combiner. Figure 3.10 describes the results of the microphone 2 located at the right side of the power combiner. Figure 3.11 depicts the noise of the microphone 3 located at the bottom of the power combiner. Figure 3.12 describes results from the acceleration sensor 1, and Fig. 3.13 shows results obtained by the acceleration sensor 2. In these figures, the horizontal axis represents the noise frequency, the left vertical axis means the test time, and the right vertical axis refers to the weighted noise sound pressure in dB. The depth of color represents the noise level.

It can be concluded from Figs. 3.6–3.8 that the noise frequency is mainly concentrated around 1244 Hz, 2488 Hz, and 699 Hz. In addition, it also can be seen from Figs. 3.9–3.11 that noise is also presented at 1655 Hz and 2892 Hz.

Figures 3.12 and 3.13 show that the hybrid synthesizing box has the largest vibration amplitude near 1244 Hz, and also has a large vibration amplitude near 699 Hz and 2488 Hz. In this condition, the speed of the load motor is 250 rpm and the meshing frequency of the gears in the CPGS can be expressed as [36]:

$$f_1 = \frac{250}{60} \times \frac{z_{diff}}{z_{m1}} \times \frac{z_{m2}}{z_{r2}} \times z_{r1}. \tag{3.1}$$

The meshing frequency of the outer ring and reducer is

$$f_2 = \frac{250}{60} \times \frac{z_{diff}}{z_{m1}} \times z_{m2}. \tag{3.2}$$

In the equation, $z_{diff} = 85, z_{m1} = 38, z_{m2} = 75, z_{r1} = 73, z_{r2} = 41$ denote the tooth number of the differential gear, reducers, the inner ring, and outer ring, submitting the tooth number into Eqs. (3.1) and (3.2) obtains

$$f_1 = 1244 \, \text{Hz}, \quad f_1 = 699 \, \text{Hz}. \tag{3.3}$$

Therefore, under this condition, the gear meshing in the CPGS causes the main noise. In addition, the gear meshing between ring and reducer is a key source. Moreover, the reason at

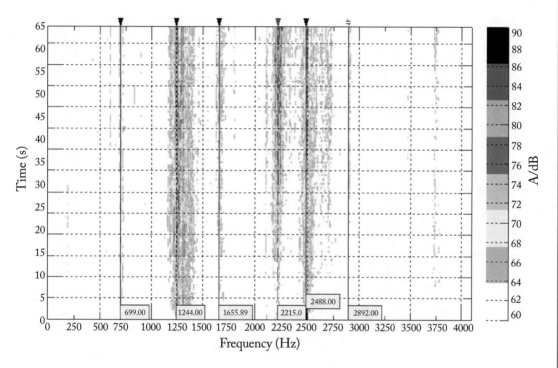

Figure 3.9: Test results for time-frequency responses of noise at microphone 1 under working condition 1 [49].

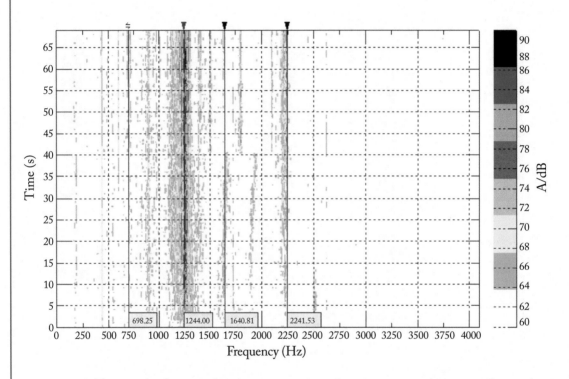

Figure 3.10: Test results for time-frequency responses of noise at microphone 2 under working condition 1.

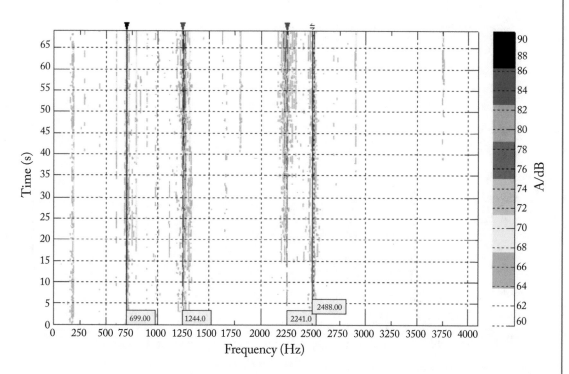

Figure 3.11: Test results for time-frequency responses of Noise at microphone 3 under working condition 1.

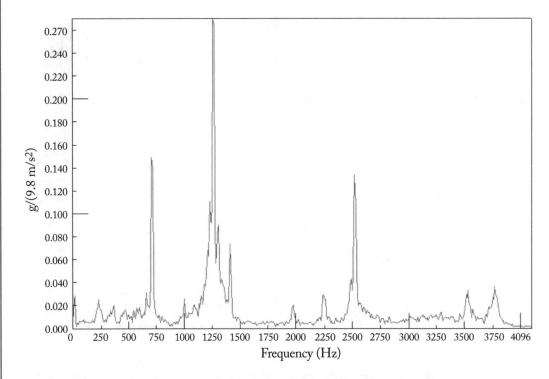

Figure 3.12: Noise test results of acceleration sensor 1 under working condition 1.

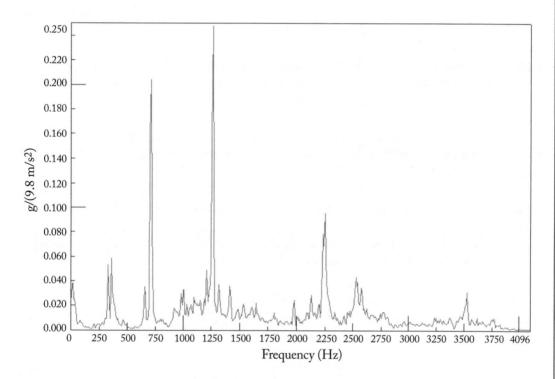

Figure 3.13: Noise test results of acceleration sensor 2 under working condition 1.

2215 Hz in Fig. 3.9 is the gear shaper cutter, which is adopted to produce the planetary gears. The noise contributions measured agree with the numerical results as well. Table 5.1 shows the 5th and 6th natural frequencies of the powertrain during the pure electric mode are 1584 Hz and 2890 Hz, respectively. Figure 3.12 shows that some noises are focus around 1656 Hz and 2886 Hz. Therefore, the conclusion is the noise contributions obtained agree with numerical ones and some mode shapes are excited in the pure driving mode.

3.3.2   TEST RESULTS OF WORKING CONDITION 2

Working condition 2 is the pure electric driving mode, where brake $B1$ is locked and the load motor speed is 350 rpm. In addition, only motor $E1$ works and its torque is -30 Nm ~ 0 Nm. Figure 3.14 indicates the results of the microphone 1 in front of the power combiner. Figure 3.15 depicts the results of the microphone 2 located at the right part of the power combiner. Figure 3.16 depicts the noise of the microphone 3 located at the bottom of the power combiner. Figure 3.17 shows the results of the acceleration sensor 1, and Fig. 3.18 shows the results of the acceleration sensor 2.

Figures 3.14–3.16 show that the noise frequency is mainly concentrated around 1742 Hz, 3144 Hz, 3488 Hz, and 978 Hz. In addition, it also can be seen from Figs. 3.11–3.13 that noise is also presented at 1572 and 2734 Hz.

Figures 3.17 and 3.18 show that the hybrid synthesizing box has the largest vibration amplitude near 699 Hz, and also has large vibration amplitude near 1742 Hz, 3144 Hz, 3500 Hz, and 2180 Hz.

In this condition, the speed of the load motor is 250 rpm, and the meshing frequency of the gears in the CPGS can be expressed as [36]:

$$f_1 = \frac{350}{60} \times \frac{z_{diff}}{z_{m1}} \times \frac{z_{m2}}{z_{r2}} \times z_{r1} = 1742 \, \text{Hz}. \tag{3.4}$$

The meshing frequency of the outer ring and reducer is:

$$f_2 = \frac{350}{60} \times \frac{z_{diff}}{z_{m1}} \times z_{m2} = 978 \, \text{Hz}. \tag{3.5}$$

Therefore, under this mode, the gear meshing cause the main noise in the CPGS and the gear meshing between outer ring and reducer is a major noise source. In addition, the contribution at 3144 Hz is the gear shaper cutter, which is utilized to produce planetary gears. The noise of 3500 Hz is the double frequency of 1742 Hz, which also agrees with the numerical results. Moreover, the 5th and 6th natural frequencies of the powertrain in pure electric mode are 1584 Hz and 2890 Hz, respectively. Figure 3.14 indicates that some noises appear around 1656 Hz and 2886 Hz. Thus, the conclusion is that noise contributions obtained agree with the numerical results and some mode shapes are excited in this pure mode.

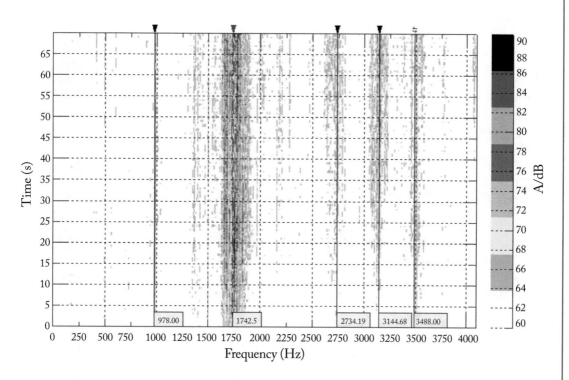

Figure 3.14: Test results for time-frequency responses of noise at microphone 1 under working condition 2.

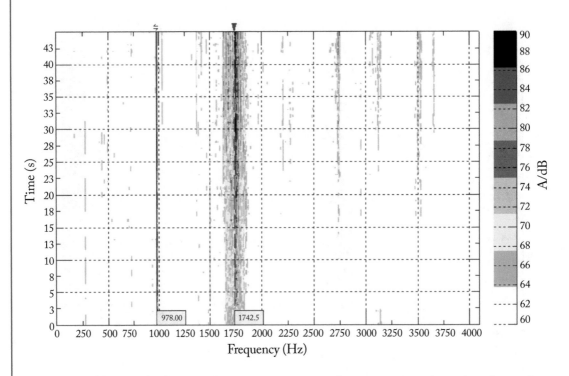

Figure 3.15: Test results for time-frequency responses of noise at microphone 2 under working condition 2.

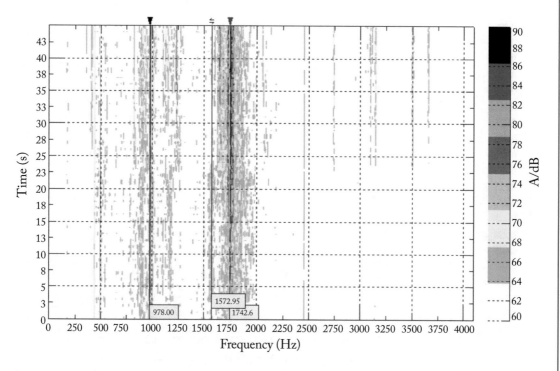

Figure 3.16: Test results for time-frequency responses of noise at microphone 3 under working condition 2.

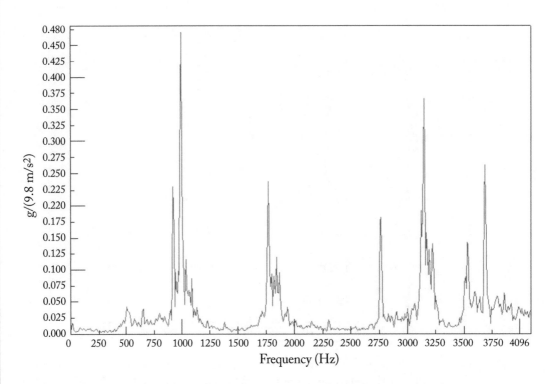

Figure 3.17: Noise test results of acceleration sensor 1 under working condition 2.

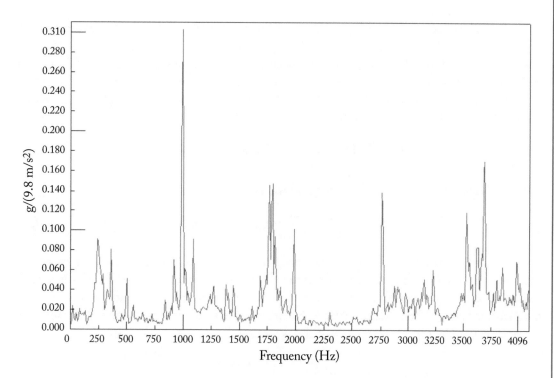

Figure 3.18: Noise test results of acceleration sensor 2 under working condition 2.

3.3.3   TEST RESULTS OF WORKING CONDITION 3

Working condition 3 is the pure electric mode, where brake $B1$ is locked and the loading motor speed is from 0–500 rpm. Only motor $E1$ works with a torque -30 Nm. Figure 3.19 indicates the results of the microphone 1 located in front of the power combiner. Figure 3.20 refers to the results of the microphone 2 located at the right part of the power combiner. Figure 3.21 depicts the noise of the microphone 3 located at the bottom of the power combiner. In these figures, the horizontal axis refers to the noise frequency, the left vertical axis shows the speed of loading motor, and the right vertical axis shows the weighted noise sound pressure in dB. The depth of color represents the noise level.

It can be concluded from Figs. 3.19–3.21 the noise is changed with the increasing speed of the loading motor. Accordingly, the noise frequency is mainly concentrated on order 298, 167, 532, and 596. In addition, it also can be seen that noise is also presented at 1600 Hz and 2800 Hz, which are corresponded to the 5th and 6th natural frequencies of the hybrid system.

Under this condition, the speed of the load motor changes from 0–250 rpm, and the meshing frequency of the gear can be expressed as [36]:

$$f_1 = \frac{n_{motor}}{60} \times \frac{z_{diff}}{z_{m1}} \times \frac{z_{m2}}{z_{r2}} \times z_{r1} = 298 \cdot n_{motor}. \tag{3.6}$$

The meshing frequency of the ring and the reducer is

$$f_1 = \frac{n_{motor}}{60} \times \frac{z_{diff}}{z_{m1}} \times z_{m2} = 167 \cdot n_{motor}. \tag{3.7}$$

According to Eqs. (3.6) and (3.7), under this work condition, gear meshing causes the main noise in the CPGS and the gear meshing between the outer ring and reducer is a major source. Moreover, the contribution at order 532 is the gear shaper cutter adopted to compose the planetary gears. The noise of order 596 is the double frequency of order 298. In addition, the noise caused by the TV resonance of the powertrain is also the main noise.

3.3.4   TEST RESULTS OF WORKING CONDITION 4

Working condition 4 is the pure electric mode, where brake $B1$ is locked and the load motor speed is from 0–500 rpm. Only motor $E2$ works and its torque is 60 Nm. Figure 3.22 presents the results of the microphone 1 located in front of the power combiner. Figure 3.23 indicates the results of the microphone 2 located at the right part of the power combiner. Figure 3.24 depicts the results from microphone 3 located at the bottom of the power combiner.

The experimental result in this mode is similar to the result of working condition 3. The noise frequency is mainly concentrated in order 298, and there are also large noises in order 167, 532, and 596. There are also strong noises around 1600 Hz and around 2800 Hz, which do not change with the changing load motor speed, and they are close to the natural frequencies of the 5th and 6th orders of the hybrid powertrain.

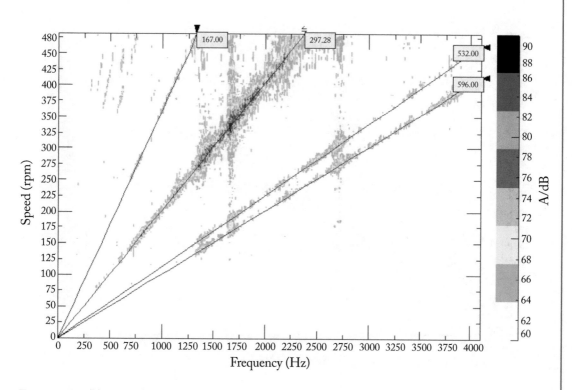

Figure 3.19: Test results for time-frequency responses of noise at microphone 1 under working condition 3.

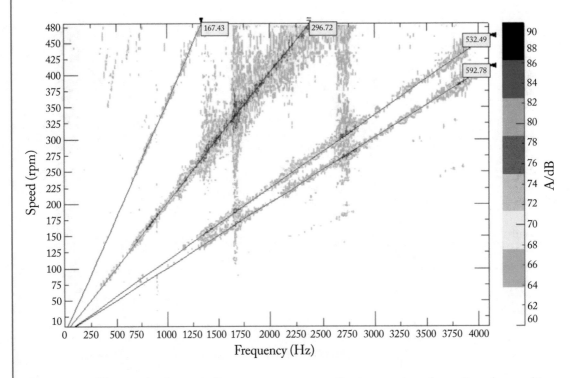

Figure 3.20: Test results for time-frequency responses of noise at microphone 2 under working condition 3.

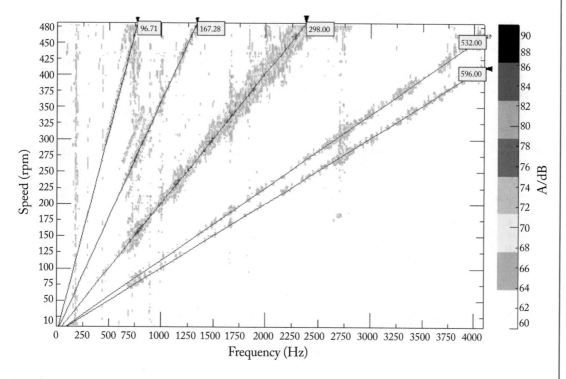

Figure 3.21: Test results for time-frequency responses of noise at microphone 3 under working condition 3.

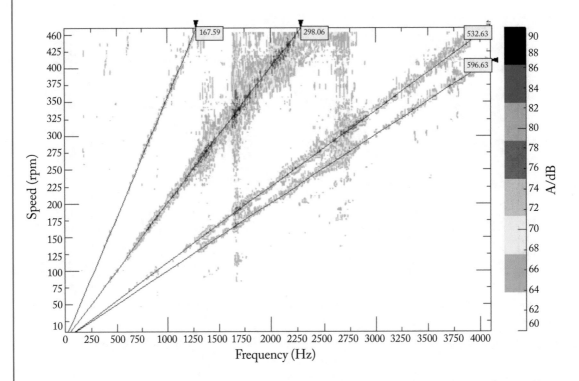

Figure 3.22: Test results for time-frequency responses of noise at microphone 1 under working condition 4.

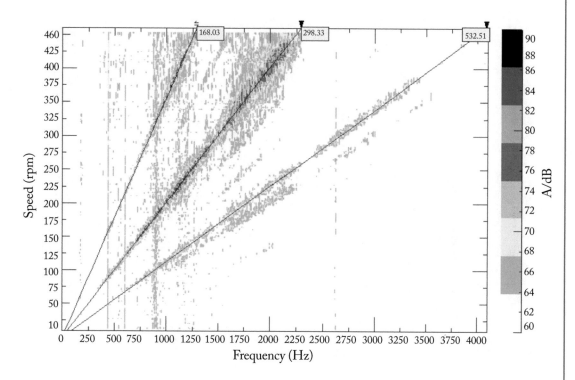

Figure 3.23: Test results for time-frequency responses of noise at microphone 2 under working condition 4.

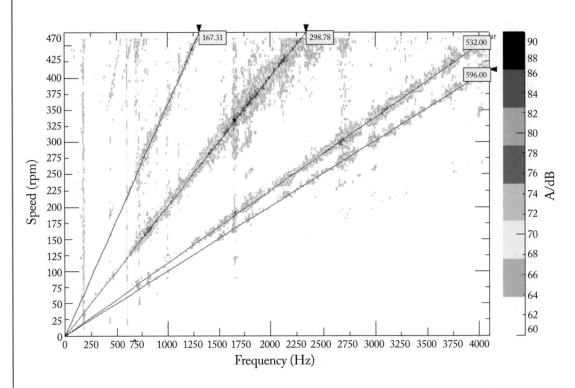

Figure 3.24: Test results for time-frequency responses of noise at microphone 3 under working condition 4.

Therefore, under the pure electric working conditions, the internal gear meshing induces the main noise the planetary and the meshing noise of the outer ring gear and the reducer is also relatively large. The noise caused by the machining of the planetary gear cutter is also large. Finally, the self-sounding noise caused by TV in the transmission system is also an important noise of a hybrid car.

3.3.5   TEST RESULTS OF WORKING CONDITION 5

TV under working condition 5 is a hybrid mode with a pedal opening of 35% and a load motor speed of 150–750 rpm. Figures 3.25 and 3.26 show the microphone noise test results at the front and right of the power synthesis box, respectively.

Under this hybrid condition, the motor drives the vehicle to start, the engine ignites when the load motor speed reaches 350 rpm, and the drivetrain power is provided by the engine and the motor after 350 rpm. According to the test results, before the engine is not ignited, the noise is mainly concentrated in order 298, order 532, order 596, and order 167. When the engine is ignited, order 298, order 532, and order 596 disappear, order 167 become the main source of the noise. Therefore, under hybrid conditions, when the engine stops, the gear meshing in the planetary row is the main noise source. The meshing of the outer ring and the reducer is the main source, and the noise caused by machining the planetary gear tool is also important; when the engine starts, the mesh noise of the ring and reducer and the engine are the main noise source of the drivetrain.

Test results obtained in different working modes are presented in Table 3.4. Compared to numerical results in Tables 5.1 and 5.2, authors conclude the order 6 of natural frequencies under pure electric driving mode is about 1610 Hz and the order 7 in Table 5.1 is relevant to 2810 Hz in condition 1. The order 5 of the natural frequency of pure electric driving mode is corresponding to 432 Hz in condition 2. While during the hybrid driving mode, the order 7 of natural frequency in Table 5.2 is corresponding to 2100 Hz in test 4. After the comparison between test results and natural frequencies of power-split hybrid driveline, it can be seen that some natural frequencies are the important noise source both in pure and hybrid driving mode conditions. Thus, self-excited is the main reason of noise in HEV driveline.

Table 3.4: Test results

Working Condition	The Order/Frequency of the Noise
1	298, 167, 96, 596; 1610 Hz, 2810 Hz
2	886 Hz, 160 Hz, 432 Hz
3	1742 Hz, 978 Hz, 3127 Hz
4	1050 Hz, 2100 Hz, the noise of the engine
5	167, the noise of the engine

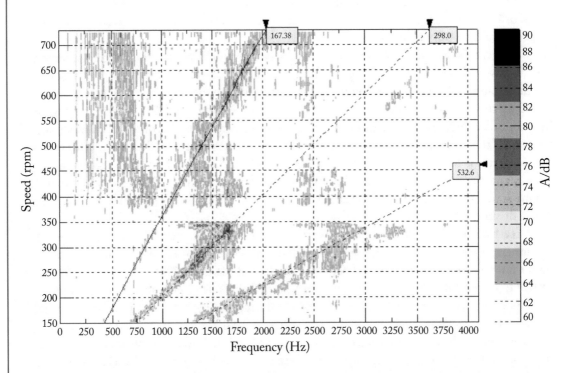

Figure 3.25: Test results for time-frequency responses of noise at microphone 1 under working condition 5.

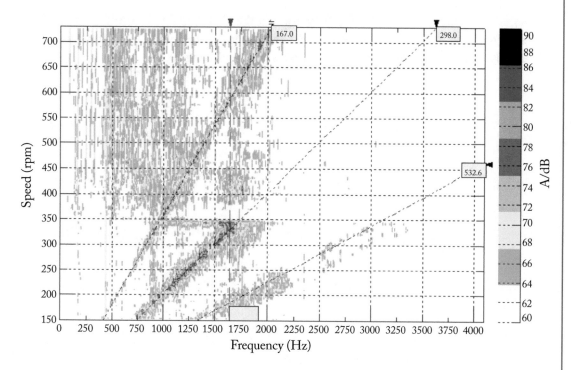

Figure 3.26: Test results for time-frequency responses of noise at microphone 2 under working condition 5.

Experimental results under pure electric driving modes show that the main noise sources of the hybrid powertrain are the meshing gear pairs in the CPGS. Then, the gear pair of ring and reducer, as well as the gear shaper cutter, are also the main noise sources. The experimental results of hybrid driving modes show that the engine is the main noise source, and the engagement between ring and reducer is always a major source of noise regardless if the engine is working or not. Furthermore, the self-excited noise is an important noise in a hybrid powertrain.

3.4   CONCLUSIONS

The noise experiments indicate the meshing noise in the CPGS and between ring and reducer is the major noise during the pure electric driving mode. In addition, it also concluded that self-exited noises contribute more during the pure electric driving mode.

CHAPTER 4

Transmission System Parameters and Meshing Stiffness Calculation

In this section, the meshing stiffness is first computed according to the Ishikawa formula [37], and the meshing noises of the gear are then derived. In this chapter, the gear width is set as b, the tooth surface normal load on the gear is F_N, the tooth width of the gear unit is ω, and the total teeth deformation is defined as δ.

Based on the above definitions, the stiffness k of the individual tooth of the gear can be defined by [37]:

$$k = \frac{F_N}{\delta \cdot b}. \tag{4.1}$$

The comprehensive meshing stiffness K of any gears pair can be described by:

$$K = \frac{k_1 \cdot k_2}{k_1 + k_2}, \tag{4.2}$$

where k_1 and k_2 are the tooth stiffnesses of both the drive and driven gears.

The gear teeth number in the hybrid powertrain is shown in Table 4.1. The inertia and material properties of the drive train components are tabulated in Table 4.2.

Table 4.1: Gear teeth quantity in the hybrid transmission

	Name	Number of Teeth	Serial Number	Name	Number of Teeth
1	Small sun gear	23	6	Outer ring gear	41
2	Big sun gear	31	7	Reducer gear	75
3	Short planetary gear	25	8	Reducer pinion	38
4	Long planetary gear	18	9	Differential gear	85
5	Inner ring gear	73			

Table 4.2: Inertia and material properties of the driveline components

	Part Name	Parameter Value
1	Engine rotating component assembly moment of inertia	0.28 kg·m²
2	Planet carrier and motor assembly moment of inertia	0.0661 kg·m²
3	Differential assembly moment of inertia	0.015 kg·m²
4	Wheel moment of inertia	1.83 kg·m²
5	Planetary frame assembly quality	2.53 kg
6	Motor 1 moment of inertia	0.027 kg·m²
7	Motor 2 moment of inertia	0.037 kg·m²
8	Differential assembly quality	4.43 kg
9	Damping shock absorber TS	618 Nm/rad
10	Left half shaft TS	5520 Nm/rad
11	Right half shaft TS	4222 Nm/rad
12	Tire TS	780 Nm/rad
13	Vehicle quality	1250 kg

Ishikawa Formula is mainly applied for gear meshing stiffness calculation. This formula regards the gear tooth end face as a combination of a rectangle and a trapezoid. The total meshing deformation is the sum of the rectangular part bending deformation, trapezoidal part deformation, shear deformation, the base part inclination deformation, and contact deformation. The graphical representation of this formula is shown in Fig. 4.1.

The deformation δ of the load that is acting along the meshing line can be calculated by the following equation [37]:

$$\delta = \delta_{Br} + \delta_{Bt} + \delta_S + \delta_G, \tag{4.3}$$

where δ_{Br} refers to the deformation of the rectangular part [37]:

$$\delta_{Br} = \frac{12 F_N \cos^2 \omega_x}{E b s_F^2} \left[h_x h_r (h_x - h_r) + \frac{h_r^3}{3} \right]. \tag{4.4}$$

δ_{Bt} is the amount of bending deformation of the trapezoidal part [37]:

$$\delta_{Bt} = \frac{6 F_N \cos^2 \omega_x}{E b s_F^3} \left[\frac{h_i - h_x}{h_i - h_r} \left\{ 4 - \frac{h_i - h_x}{h_i - h_r} \right\} - 2 \ln \frac{h_i - h_x}{h_i - h_r} - 3 \right] (h_i - h_r)^3. \tag{4.5}$$

δ_S is the amount of deformation caused by shear:

$$\delta_{Br} = \frac{12 F_N \cos^2 \omega_x}{E b s_F^2} \left[h_x h_r (h_x - h_r) + \frac{h_r^3}{3} \right]. \tag{4.6}$$

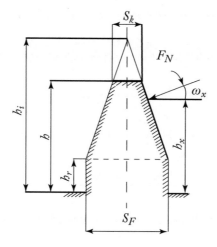

Figure 4.1: Ishikawa formula.

δ_G is the amount of deformation caused by the inclination of the base part:

$$\delta_G = \frac{24 F_N h_x^2 \cos^2 \omega_x}{\pi E b s_F^2}. \tag{4.7}$$

When a pair of teeth is engaged, the sum of the deformation amounts along the meshing line is:

$$\delta = \delta_1 + \delta_2 + \delta_{pv}, \tag{4.8}$$

where δ_1 and δ_2 are the δ values of each tooth, and δ_{pv} is the amount of deformation of the contact surface of the tooth surface.

$$\delta_{pv} = \frac{4(1 - v^2)}{\pi E} \frac{F_N}{b}. \tag{4.9}$$

According to the Ishikawa Formula, the average meshing stiffness of the gear pair in the hybrid powertrain is obtained in Matlab, and the results are shown in Table 4.3.

4.1   GEAR PAIR MESHING IMPACT RESPONSE ANALYSIS

The characteristic of the gear transmission is that the gear teeth alternately mesh with each other, and there are both rolling and sliding in the process, thus generating friction and impact between the teeth and the teeth, thereby generating vibration and causing noise [58–64]. The meshing impact of the gear is the primary cause of gear noise. The noise generated by the gear meshing impact is divided into acceleration noise and self-sounding noise. On the one hand, when the gears mesh, the impact causes the gear to generate a significant acceleration and disturb the surrounding components. The noise generated by this disturbance then becomes the acceleration

Table 4.3: Meshing stiffness of the gear pairs in the hybrid driveline

	Gear Pair	Meshing Stiffness (N/mm)
1	Short planetary gear – inner ring gear	3.11e5
2	Short planetary gear – long planetary gear	2.54e5
3	Short planet gear – small sun gear	2.64e5
4	Long planetary gear – big sun gear	2.40e5
5	Outer ring gear – large reducer gear	3.44e5
6	Small reducer gear – differential gear	6.18e5

noise of the gear. On the other hand, under the excitation of the dynamic meshing force of the gear, the components of the system will induce vibration, and the noise generated by these vibrations is called self-noise noise.

During the actual meshing of the gear, due to the gear error and the deformation of the gear teeth, the gear is at the point of mesh merging (out) away from the theoretical meshing line, which causes the driving gear and the driven gear to deviate and mutation, causing gears to bite and result in the impact. The meshing impact is the main reason for the gear transmission error, and it is also one of the primary dynamic excitations in the study of gear vibration problems.

In the meshing process of the gear pair, the deviation of the actual base and the theoretical base caused by the error of the gear and the gear load deformation is defined as the meshing base error. The synthetic base error Δt_{oc} of the driving gear and the driven gear, which can be expressed as:

$$\Delta t_{oc} = \Delta f_p - \Delta f_g, \tag{4.10}$$

where Δf_p and Δf_g are the base section error of the drive gear and the driven gear.

The meshing impact has different forms. When $\Delta t_{oc} < 0$, the pair of gears will have an impact at the moment of entering the engagement, which is known as a biting impact. When $\Delta t_{oc} > 0$, the pair of gears will create a piercing impact at the moment of exiting the engagement. The above two kinds of meshing impacts will result in huge influence on the transition process, and such influence are in the following two aspects.

1. Elastic deformation effect.

   The loaded elastic deformation of the teeth causes the actual meshing base of the driving wheel to decrease, while the actual meshing base of the driven wheel becomes larger. Thus, when $\Delta t_{oc} < 0$, the elastic deformation of the gear increases the synthetic meshing base error, so that the gearing impact of the gear is going to increase. When $\Delta t_{oc} > 0$, the elastic deformation makes the meshing base error becomes smaller and tends to be smoother, so that the impact shock becomes slighter.

2. Inertia effect.

When $\Delta t_{oc} < 0$, the impact occurs when the subsequent teeth of the driven gear just enter the meshing state. At this time, the driven gear must maintain the original speed, and the driving gear should increase its speed, which makes the meshing shock increase. When $\Delta t_{oc} > 0$, the speed of the passive gear slows down due to the inertia of the driven gear, which result in the impact of both the driving and the driven wheels speed on the meshing line becomes small.

For these reasons, the impact of the mesh merging is more significant than that of the mesh-out, so that the following discussion will focus on the impact while the gear mesh merging.

4.1.1 CALCULATION OF IMPACT ACCELERATION AND IMPACT TIME

Due to the error and elastic deformation of the gear, an impact will occur during the gear meshing process, and the gear will enter the meshing point outside the meshing line. At this time, the two tooth profiles have no common normal at the contact point (as shown in Fig. 4.2). The difference $v_a = v_{1a} - v_{2a}$ is called the impact velocity, which represents the velocity component perpendicular to the tooth profile of the driving wheel, v_a can be calculated by [37]:

$$v_a = \omega_1 r_{g1} \left(1 + \frac{1}{i}\right) \left[1 - \frac{\cos\left(\alpha_{E_1'} + \gamma_1\right)}{\cos \alpha_b}\right], \tag{4.11}$$

where ω_1 is the drive wheel angular velocity, i represents the gear ratio, α_b stands for the pressure angle on the index circle, $\alpha_{E_1'} + \gamma_1$ is the argument. When two gears M_1 and M_2 meet with speed difference, they will hit each other. Its greatest impact is:

$$F_m = v_a \cdot \sqrt{\frac{b}{q_{E_1}}} \cdot \sqrt{\frac{J_1 \cdot J_2}{J_1 r_{g2}'^2 + J_2 r_{g1}^2}}. \tag{4.12}$$

In this formula, q_{E_1} is the flexibility for engaging gear pairs, J_1, J_2 is the moment of inertia for gears, b is the tooth width, r_{g1} is the basic circle radius of the driving gear, r_{g2}' is the equivalent base circle radius of a passive gear.

If the change of impact force is assumed to be a half-wave sinusoidal pulse, the impact force of gears 1 and 2 (for convenience, the subscripts 1 and 2 are omitted in the following expressions, except as specified in particular) [37]:

$$f_a(t) = F_m \sin \omega_c t \ \ (0 \leq t \leq t_c), \tag{4.13}$$

where ω_c is the frequency of the half wave sine pulse, t_c is the impact time, then the impact acceleration can be expressed as:

$$a_a(t) = \frac{F_m}{M} \sin \omega_c t = a_m \sin \omega_c t \ \ (0 \leq t \leq t_c), \tag{4.14}$$

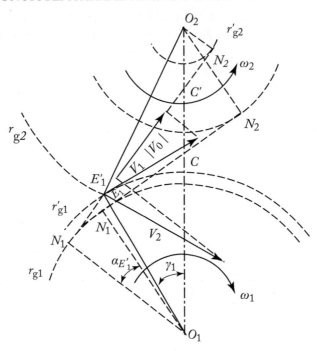

Figure 4.2: Impact velocity of meshing resulting from gear error.

where $a_m = \frac{F_m}{M}$.

Because of the errors and elastic deformation of the gears, the gear runs unsteadily, so the speed v_1 and v_2 of the two gears are unknown at the beginning of the impact, but the impact speed of the two gears can be obtained from Eq. (4.16). In order to estimate the meshing impact strength of a pair of gears, the meshing impact system of a pair of gears is simplified as follows. For the driving gear, the relative velocities of the two teeth along the tooth normal direction at the meshing point at the beginning and the end of the impact can be described by:

$$v_a = v_{1a} - v_{2a} \tag{4.15}$$
$$\Delta v = v'_{1a} - v'_{2a}. \tag{4.16}$$

The motion of the two gear should satisfy the following momentum conservation equation:

$$M_1 v_{1a} + M_2 v_{2a} = M_1 v'_{1a} + M_2 v'_{2a}. \tag{4.17}$$

According to the impulse theorem, gear 2 satisfies

$$\int_0^{t_c} f_a(t)dt = M_2 v'_{2a} - M_2 v_{2a}. \tag{4.18}$$

Considering $v_{1a} = v_{2a} + v_a, v'_{1a} = v'_{2a} + \Delta v$, Eq. (4.17) can be written as

$$(M_1 + M_2) v_{2a} - (M_1 + M_2) v_{2a} = M_1 v_a - M_1 \Delta v. \tag{4.19}$$

Denote $M = \frac{M_1 M_2}{M_1 + M_2}$ as the equivalent mass of the gear system, one can obtain

$$\int_0^{t_c} f_a(t) dt = M v_a - M \Delta v. \tag{4.20}$$

Assuming that the two gear enters normal meshing after the impact, then the relative speed of the meshing point Δv will be 0, which means

$$\int_0^{t_c} f_a(t) dt = M v_a. \tag{4.21}$$

Replace Eq. (4.13) and $\omega_c = \pi / t_c$ in the type (4.21), we can obtain

$$\int_0^{t_c} f_a(t) dt = \int_0^{t_c} F_m \sin \omega_c t \, dt = \frac{2 F_m}{\omega_c} = \frac{2 F_m}{\pi} t_c. \tag{4.22}$$

Then, the time of gear meshing impact t_c can be expressed as following:

$$t_c = \frac{\pi}{2} \cdot \frac{M v_a}{F_m}. \tag{4.23}$$

4.1.2 CALCULATION OF IMPACT ACCELERATION SOUND PRESSURE

The acceleration noise produced by gear impacting at a certain point in the process of gear grinding can be viewed as the acceleration noise produced by two cylinders with variable radius of curvature impacting each other. Since the near sound field is difficult to be calculated, We compute the sound pressure of the far sound field as [37]:

When $0 \leq t' \leq t_c$,

$$p(r, \theta, t) = A \left(B \cos \omega_c t' + D \sin \omega_c t' + E \cos l_1 t' e^{-t_2 t'} + F_1 \sin l_1 t' e^{-t_2 t'} \right). \tag{4.24}$$

When $t' > t_c$,

$$\begin{aligned} p(r, \theta, t) = A \Big\{ & \left[-G \cos \left((\omega_c + l_1) t_c - l_1 t' \right) - H \sin \left((\omega_c + l_1) t_c - l_1 t' \right) \right. \\ & \left. + X \cos \left((\omega_c - l_1) t_c - l_1 t' \right) + Y \sin \left((\omega_c - l_1) t_c - l_1 t' \right) \right] e^{-t_2 (t' - t_c)} \\ & + \left[E \cos l_1 t' + F \sin l_1 t' \right] e^{-t_2 t'} \Big\}, \end{aligned} \tag{4.25}$$

where $l_1 = \frac{\sqrt{95}c}{16a}$, $l_2 = \frac{7c}{16a}$, $t' = t - \frac{r-a}{c}$, and c is the velocity of sound.

$$A = \sqrt{\frac{a}{r}} \cdot \frac{a^2 \cos\theta}{r} a_m \rho_0$$

$$B = \frac{1}{\sqrt{95}} \left(\frac{8r}{a} - \frac{7}{3} \right) (-A_1 + B_1)\omega_c - \frac{1}{3}(C_1 + D_1)\omega_c$$

$$D = 1 + \frac{1}{\sqrt{95}} \left(\frac{8r}{a} - \frac{7}{3} \right) (C_1 - D_1)\omega_c - \frac{1}{3}(A_1 + B_1)\omega_c$$

$$E = \frac{1}{\sqrt{95}} \left(\frac{8r}{a} - \frac{7}{3} \right) (E_1 + F_1) + \frac{1}{3}(C_1 + D_1)\omega_c$$

$$F = \frac{1}{\sqrt{95}} \left(\frac{8r}{a} - \frac{7}{3} \right) (-C_1 - D_1)\omega_c + \frac{1}{3}(E_1 + F_1)$$

$$G = \frac{1}{\sqrt{95}} \left(\frac{8r}{a} - \frac{7}{3} \right) F_1 + \frac{1}{3}D_1\omega_c$$

$$H = \frac{1}{\sqrt{95}} \left(\frac{8r}{a} - \frac{7}{3} \right) D_1\omega_c - \frac{1}{3}F_1$$

$$X = -\frac{1}{\sqrt{95}} \left(\frac{8r}{a} - \frac{7}{3} \right) E_1 - \frac{1}{3}C_1\omega_c$$

$$Y = \frac{1}{\sqrt{95}} \left(\frac{8r}{a} - \frac{7}{3} \right) C_1\omega_c - \frac{1}{3}E_1.$$

Inside,

$$A_1 = \frac{\omega_c - l_1}{(\omega_c - l_1)^2 + l_2^2}$$

$$B_1 = \frac{\omega_c + l_1}{(\omega_c + l_1)^2 + l_2^2}$$

$$C_1 = \frac{l_2}{(\omega_c - l_1)^2 + l_2^2}$$

$$D_1 = \frac{l_2}{(\omega_c + l_1)^2 + l_2^2}$$

$$E_1 = \frac{\omega_c l_1 - l_1^2 - l_2^2}{(\omega_c - l_1)^2 + l_2^2}$$

$$F_1 = \frac{\omega_c l_1 + l_1^2 + l_2^2}{(\omega_c + l_1)^2 + l_2^2}.$$

r in the formula represents the distance from the point of space to the curvature center of the tooth profile, as shown in Fig. 4.2, $r_1 = ON_1, r_2 = ON_2$, which can be computed as per the geometric relationship of Fig. 4.3. As long as the test distance r is large enough relative to the gear tooth width b, it can be regarded as far-field test. a is the radius of curvature of the gear meshing impact point, as shown in Fig. 4.3. $a_1 = N_1 B_2$ and $a_2 = N_2 B_2$ can be calculated from the meshing principle.

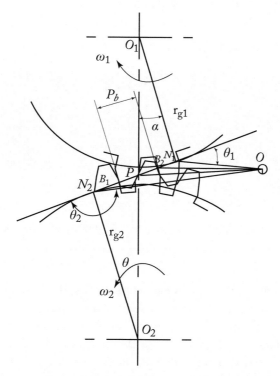

Figure 4.3: Position parameter of gear pair under the condition of meshing.

The sound pressure of a meshing gear system at r points in space can be expressed as follows [37]:

$$p(r, \theta, t) = p_1(r_1, \theta_1, t) + p_2(r_2, \theta_2, t), \qquad (4.26)$$

where $\theta_1 = \arcsin \frac{OP \cdot \sin a_b}{ON_1}$, $\theta_2 = \pi = -\arcsin \frac{OP \cdot \sin a_b}{ON_2}$.

4.2   ANALYSIS OF RESULTS

Through the theories above, the gear meshing noise can now be calculated. In this part, the gear meshing pair in the composite planetary row and the drive train is regarded as the object. First, meshing noise of the two gears in each gear pair has the circumferential section error and the tooth profile error. Taking the small sun gear the planetary gear as an example, the effect of the circumferential section error, the tooth profile error, the meshing stiffness and the driving gear speed on the meshing noise of the gear pair will be calculated.

Table 4.4 shows the meshing impact noise of the gear pair in the compound planetary row and the hybrid powertrain with the error (circumference error and tooth profile error). The gear pitch error is 2×10^{-2} mm, the tooth profile error is 2×10^{-2} mm, the meshing stiffness is 2×10^7 N/m, and the drive gear speed is 200 rad/s.

Table 4.4: Meshing noise of gear pairs

Gear Pair	Short Planet Gear - Small Sun Gear	Short Planetary Gear - Long Planetary Gear	Long Planet Gear - Big Sun Gear	Reducer Gear - Ring Gear	Reducer Gear- differential
Gear noise when the former has error (Pa)	1.40×10^{-2}	1.39×10^{-2}	1.14×10^{-2}	1.31×10^{-2}	3.69×10^{-3}
Gear noise when the latter has error (Pa)	7.01×10^{-4}	7.75×10^{-3}	7.07×10^{-5}	3.43×10^{-3}	8.21×10^{-3}

As can be seen from Table 4.4, in a pair of gear pairs, the induced meshing noise varied depending on the sources. When the long and short planet wheels are meshed, the error of the short planet gear has significant influence on the impact noise, and when the short planetary gear meshes with the small sun gear, the error of the short planetary gear has a more significant influence on the impact noise. When the long planetary gear meshes with the large sun gear, the long planetary gear error contributes significantly to the impact noise, and when the reducer gear meshes with the ring gear, the error of the reducer gear has a greater impact on the impact noise. When the reducer gear meshes with the ring gear, the ring gear error can cause considerable impact noise. When the reducer gear meshes with the differential, the differential error has a significant influence on the meshing noise.

It can also be seen from Table 4.4 that the noise of the short planetary gear in the planetary row is more significant for the meshing impact, followed by the long planetary gear. The small sun gear has less influence on the impact noise, and when the gear of the reducer meshes with the ring gear, the influence of the error on the noise is greater than the influence of the error of the external gear of the ring gear on the noise. When the gear of the reducer meshes with the gear of the differential, the differential gear error has greater effect on the noise.

4.3   CONCLUSIONS

This chapter tested the gantry noise and vibration for the hybrid powertrain. The tests were performed in the powertrain laboratory of Geely Automobile Co., Ltd. In the test, three sound pressure sensors were mounted to the front end, the right end and the lower end of the power synthesis box. In addition, a number of accelerators were complicated to test the acceleration response amplitude of the drive train. A photoelectric speed sensor was used to follow the load motor speed. The experiment tested included five typical test conditions, i.e., four pure electric conditions and one hybrid operating condition.

Through analysis and test results, it can be concluded that in pure electric mode, the meshing noise of the inner planetary gear in the power synthesis box was the dominant one, and the meshing of the outer ring gear as well as the reducer gear also played an important role. Besides, the noise caused by cutting tools from planetary gears and the self-sounding noise caused by the TV were also an important noise. As for the hybrid mode, when the engine was not started, the noise of the planetary gear internal meshing was the primary source, and the outer ring gear as well as the gear noise of the reducer and the noise caused by the planetary gear machining tool were important noises of the drive train. When the engine was ignited, the outer ring gear and the reducer gear meshing noise and engine noise became the main noise.

The gear meshing noise in the compound planetary row was the main noise of the hybrid powertrain in the pure electric mode. Therefore, it was obliged to analyze the gear pair meshing noise in the compound planetary row to find the gear meshing pair that produced the most noise, considering different factors on the gear meshing noise, and optimize the tooth profile to reduce the meshing noise of the inner gear of the composite planetary row. In order to reduce the noise and TV of hybrid vehicles, it was necessary to model the hybrid vehicle and perform TV analysis to reduce the TV and noise of the hybrid powertrain and propose improvements.

CHAPTER 5

Mathematical Modeling and TV Analysis of Hybrid Electric Vehicles

The TV of the powertrain is one of the primary sources for vibration and noise of the vehicle, which has a significant impact on the vehicle NVH performance. In order to reduce the TV and noise of the hybrid vehicle, it is significantly necessary to carry out the TV modeling and analysis of hybrid vehicle drive train [65–72].

The automotive powertrain is a complex TV system and an important source of vehicle NVH. The TV of the drive train is an important form of vibration of the vehicle. It is coupled with vehicle longitudinal vibration, the TV of the axle suspension system, and the bending vibration of the drive train, which result in complex vibration and noise problems. It is thus of great theoretical and practical significance to study the TV of automotive transmission systems [73–80].

The most significant difference between a hybrid vehicle and a conventional ICE vehicle appears in the transmission system. Since the hybrid vehicle has multiple power sources, a power coupling device is used to couple multiple powers, and the power of each power source is allocated through control. The power coupling is part of the core of a hybrid vehicle, and its performance directly affects the performance of the hybrid vehicle's power, vibration, noise, and comfort. The research carried out in this part is based on the compound planetary row deep hybrid vehicle developed by Geely Automobile Co., Ltd. The powertrain structure mainly includes engine, TV damper, compound planetary row device, two motors, reducer, differential, half shaft, and the wheel. The compound planetary platoon is the power coupling and power splitting device of the hybrid vehicle, which is the core part of the hybrid powertrain and is a mutated Ravigneaux structure. After the engine and the dual motor are coupled by the compound planetary slab, power is transmitted to the wheels via the ring gear, reducer, differential, and half shaft to drive the car. When the excitation frequency of the engine or motor is equal to the natural frequency of the TV of the drive train, the drive train will resonate, and the force of the relevant components, especially the composite planetary row, will increase significantly, and also directly affect the NVH of the hybrid vehicle and comfort performance. Therefore, it is necessary to perform TV analysis of the composite planetary power split-type hybrid vehicle.

The full hybrid vehicle has multiple power sources and is a complex electromechanical coupling system. The establishment of the TV model is more complicated than the traditional automobile, especially the establishment of the dynamic coupling device-composite planetary TV model. To solve this problem, the dynamic equation of the compound planetary gear train is deduced by Lagrangian equation, and the dynamic equations of the components of the hybrid powertrain are listed to establish the TV model of hybrid powertrain, and the TV mode of the powertrain are calculated.

5.1 DYNAMIC MODELING OF THE COMPOUND PLANETARY GEAR SET

Figure 5.1 shows the dynamic model of the compound planetary gear set [49].

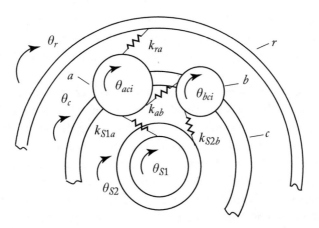

Figure 5.1: Dynamic model of the compound planetary gear set.

In Fig. 5.1 k_{S2b} is the meshing stiffness between big sun gear and long planet, k_{ab} represents the meshing stiffness between short planet and long planet, k_{S1a} is the meshing stiffness between small sun gear and short planet, k_{ar} is the meshing stiffness between ring and short planet, θ_{S1} is the angular displacement of small sun gear, θ_{S2} is the angular displacement of big sun gear, θ_r is the angular displacement of ring gear, θ_c is the angular displacement of carrier, θ_{aci} and θ_{bci} are the angular displacement of short and long planet, respectively.

The Lagrangian equation of the torsional dynamic system can be described by [72]:

$$L = \frac{1}{2}I_c\dot{\theta}_c^2 + \frac{1}{2}nI_a\left(\dot{\theta}_{aci} + \dot{\theta}_c\right)^2 + \frac{1}{2}nI_a\left(\dot{\theta}_{bci} + \dot{\theta}_c\right)^2 + \frac{1}{2}I_{S1}\dot{\theta}_{S1}^2 + \frac{1}{2}I_{S2}\dot{\theta}_{S2}^2 + \frac{1}{2}I_r\dot{\theta}_r^2$$

$$- \frac{1}{2}k_{S1a}\sum_{i=1}^{3}[r_a\left(\theta_{aci} + \theta_c\right) - r_{ca}\theta_c + r_{S1}\theta_{S1}]^2$$

$$- \frac{1}{2}k_{S2b}\sum_{i=1}^{3}[r_b\left(\theta_{bci} + \theta_c\right) - r_{cb}\theta_c + r_{S2}\theta_{S2}]^2$$

$$- \frac{1}{2}k_{ar}\sum_{i=1}^{3}[r_a\left(\theta_{aci} + \theta_c\right) + r_{cb}\theta_c - r_r\theta_r]^2 - \frac{1}{2}k_{ab}\sum_{i=1}^{3}[r_a\theta_{aci} + r_b\theta_{bci}]^2. \quad (5.1)$$

The equations can be expressed by the generalized coordinates as follows:

$$\frac{d}{dt}\left(\frac{\partial L}{\partial \dot{\theta}_q}\right) - \frac{\partial L}{\partial \theta_q} = 0, \qquad q = c, S1, S2, r, ac1, ac2, ac3, bc1, bc2, bc3, \quad (5.2)$$

where the derivate of a variable represents differentiation w.r.t. time. By combing Eqs. (5.1) and (5.2), we can rewrite the coupled homogeneous ordinary differential equations as follows:

$$(I_c + nI_a + nI_b)\ddot{\theta}_c + \sum_{j=1}^{3}I_a\ddot{\theta}_{cai} + \sum_{j=1}^{3}I_b\ddot{\theta}_{cbi} - k_{S1a}r_{S1}\sum_{j=1}^{3}[r_a\theta_{cai} + r_{S1}\theta_{S1} - r_{S1}\theta_c]$$

$$- k_{S2b}r_{S2}\sum_{i=1}^{3}[r_b\theta_{cbi} + r_{S2}\theta_{S2} - r_{S2}\theta_c] + k_{ar}r_r\sum_{i=1}^{3}[r_a\theta_{cai} + r_r\theta_c - r_r\theta_r] = 0 \quad (5.3)$$

$$I_{S1}\ddot{\theta}_{S1} + k_{S1a}r_{S1}\sum_{i=1}^{3}[r_a\theta_{cai} - r_{S1}\theta_c + r_{S1}\theta_{S1}] = 0 \quad (5.4)$$

$$I_{S2}\ddot{\theta}_{S2} + k_{S2b}r_{S2}\sum_{i=1}^{3}[r_b\theta_{cbi} - r_{S2}\theta_c + r_{S2}\theta_{S2}] = 0 \quad (5.5)$$

$$I_r\ddot{\theta}_r - k_{ar}r_r\sum_{i=1}^{3}[r_a\theta_{cai} + r_r\theta_c - r_r\theta_r] = 0 \quad (5.6)$$

$$I_a\ddot{\theta}_{ai} + I_a\ddot{\theta}_c + k_{S1a}r_a[r_a\theta_{cai} - r_{S1}\theta_c + r_{S1}\theta_{S1}] + k_{ab}r_a[r_a\theta_{cai} + r_b\theta_{cbi}]$$
$$+ k_{ab}r_a[r_a\theta_{cai} + r_r\theta_c - r_r\theta_r] = 0, \qquad i = 1, 2, 3 \quad (5.7)$$

$$I_b\ddot{\theta}_{bi} + I_b\ddot{\theta}_c + k_{S2b}r_b\left[r_b\theta_{cbi} - r_{S2}\theta_c + r_{S2}\theta_{S2}\right]$$
$$+ k_{ab}r_b\left[r_a\theta_{cai} + r_b\theta_{cbi}\right] = 0, \qquad i = 1, 2, 3. \tag{5.8}$$

The equilibrium equation of the compound planetary gear set TV can be described by:

$$M'\ddot{q} + K'q = 0, \tag{5.9}$$

where the displacement vector $\{q\}$, the mass matrice $[M']$ and stiffness matrice $[K']$ are given as follows:

$$\{q\} = \begin{bmatrix} \theta_c & \theta_{S1} & \theta_{S2} & \theta_r & \theta_{ac1} & \theta_{ac2} & \theta_{ac3} & \theta_{bc1} & \theta_{bc2} & \theta_{bc3} \end{bmatrix}^T \tag{5.10}$$

$$[M'] = \begin{bmatrix}
J_c + 3J_a + 3J_b & 0 & 0 & 0 & J_{a1} & J_{a2} & J_{a3} & J_{b1} & J_{b2} & J_{b3} \\
0 & J_{S1} & 0 & 0 & 0 & 0 & 0 & 0 & 0 & 0 \\
0 & 0 & J_{S2} & 0 & 0 & 0 & 0 & 0 & 0 & 0 \\
0 & 0 & 0 & J_r & 0 & 0 & 0 & 0 & 0 & 0 \\
J_{a1} & 0 & 0 & 0 & J_{a1} & 0 & 0 & 0 & 0 & 0 \\
J_{a2} & 0 & 0 & 0 & 0 & J_{a2} & 0 & 0 & 0 & 0 \\
J_{a3} & 0 & 0 & 0 & 0 & 0 & J_{a3} & 0 & 0 & 0 \\
J_{b1} & 0 & 0 & 0 & 0 & 0 & 0 & J_{b1} & 0 & 0 \\
J_{b2} & 0 & 0 & 0 & 0 & 0 & 0 & 0 & J_{b2} & 0 \\
J_{b3} & 0 & 0 & 0 & 0 & 0 & 0 & 0 & 0 & J_{b3}
\end{bmatrix} \tag{5.11}$$

$$[K'] = \begin{bmatrix}
k'_{11} & k'_{12} & k'_{13} & k'_{14} & k'_{15} & k'_{16} & k'_{17} & k'_{18} & k'_{19} & k'_{1,10} \\
k'_{21} & k'_{22} & 0 & 0 & 0 & 0 & 0 & 0 & 0 & 0 \\
k'_{31} & 0 & k'_{33} & 0 & 0 & 0 & 0 & 0 & 0 & 0 \\
k'_{41} & 0 & 0 & k'_{44} & 0 & 0 & 0 & 0 & 0 & 0 \\
k'_{51} & 0 & 0 & 0 & k'_{55} & 0 & 0 & k'_{58} & 0 & 0 \\
k'_{61} & 0 & 0 & 0 & 0 & k'_{66} & 0 & 0 & k'_{69} & 0 \\
k'_{71} & 0 & 0 & 0 & 0 & 0 & k'_{77} & 0 & 0 & k'_{7,10} \\
k'_{81} & 0 & 0 & 0 & k'_{85} & 0 & 0 & k'_{88} & 0 & 0 \\
k'_{91} & 0 & 0 & 0 & 0 & k'_{96} & 0 & 0 & k'_{99} & 0 \\
k'_{10,1} & 0 & 0 & 0 & 0 & 0 & k'_{10,7} & 0 & 0 & k'_{10,10}
\end{bmatrix} \tag{5.12}$$

where

$$k'_{11} = 3k_{S1a}r_{S1}^2 + 3k_{S2b}r_{S2}^2 + 3k_{ar}r_r^2 \tag{5.13}$$

$$k'_{12} = -3k_{S1a}r_{S1}^2 \tag{5.14}$$

$$k'_{13} = -3k_{S2b}r_{S2}^2 \tag{5.15}$$

$$k'_{14} = -3k_{ar}r_r^2 \tag{5.16}$$

$$k'_{15} = k_{ar}r_r r_a - k_{S1a}r_{S1}r_a \tag{5.17}$$

$$k'_{16} = k_{ar}r_r r_a - k_{S1a}r_{S1}r \tag{5.18}$$

$$k'_{17} = k_{ar}r_r r_a - k_{S1a}r_{S1}r_a \tag{5.19}$$

$$k'_{18} = -k_{S2b}r_{S2}r_b \tag{5.20}$$

$$k'_{19} = -k_{S2b}r_{S2}r \tag{5.21}$$

$$k'_{1,10} = -k_{S2b}r_{S2}r_b \tag{5.22}$$

$$k'_{21} = -3k_{S1a}r_{S1}^2 \tag{5.23}$$

$$k'_{22} = 3k_{S1a}r_{S1}^2 \tag{5.24}$$

$$k'_{31} = -3k_{S2b}r_{S2}^2 \tag{5.25}$$

$$k'_{33} = 3k_{S2b}r_{S2}^2 \tag{5.26}$$

$$k'_{41} = 3k_{S2b}r_{S2}^2 \tag{5.27}$$

$$k'_{44} = 3k_{ar}r_r^2 \tag{5.28}$$

$$k'_{51} = k_{ar}r_r r_a - k_{S1a}r_{S1}r_a \tag{5.29}$$

$$k'_{55} = r_a^2(k_{S1a} + k_{ab} + k_{ar}) \tag{5.30}$$

$$k'_{58} = k_{ab}r_a r_b \tag{5.31}$$

$$k'_{61} = k_{ar}r_r r_a - k_{S1a}r_{S1}r_a \tag{5.32}$$

$$k'_{66} = r_a^2(k_{S1a} + k_{ab} + k_{ar}) \tag{5.33}$$

$$k'_{69} = k_{ab}r_a r_b \tag{5.34}$$

$$k'_{71} = k_{ar}r_r r_a - k_{S1a}r_{S1}r_a \tag{5.35}$$

$$k'_{77} = r_a^2(k_{S1a} + k_{ab} + k_{ar}) \tag{5.36}$$

$$k'_{7,10} = k_{ab}r_a r_b \tag{5.37}$$

$$k'_{81} = -k_{S2b}r_{S2}r_b \tag{5.38}$$

$$k'_{85} = k_{ab}r_a r_b \tag{5.39}$$

$$k'_{58} = k_{ab}r_a r_b \tag{5.40}$$

$$k'_{88} = k_{S2b}r_b^2 + k_{ab}r_b^2 \tag{5.41}$$

$$k'_{91} = -k_{S2b}r_{S2}r_b \tag{5.42}$$

$$k'_{96} = k_{ab}r_a r_b \tag{5.43}$$

$$k'_{99} = k_{S2b}r_b^2 + k_{ab}r_b^2 \tag{5.44}$$

$$k'_{10,1} = -k_{S2b}r_{S2}r_b \tag{5.45}$$

$$k'_{10,7} = k_{ab}r_a r_b \tag{5.46}$$

$$k'_{10,10} = k_{S2b}r_b^2 + k_{ab}r_b^2 \tag{5.47}$$

$$J_c = J'_c + n(M_a r_{ca}^2 + M_b r_{cb}^2) \tag{5.48}$$

$$r_{ca} = r_{S1} + r_a \tag{5.49}$$

$$r_{cb} = r_{S2} + r_b \tag{5.50}$$

where the moment of inertia of the carrier J_c is defined as

$$J_c = J'_c + n(M_a r_{ca}^2 + M_b r_{cb}^2), \quad r_{ca} = r_{S1} + r_a \quad r_{cb} = r_{S2} + r_b, \tag{5.51}$$

where J'_c is the carrier moment of inertia, n is total number of planet gears a-b in the compound planetary gear set, M_a and M_b are masses of planets a and b, r_{ca} and r_{cb} are radiuses of revolution of the short and long planet gears. J_{S1} is the moment of inertia of small sun gear, J_{S2} is the moment of inertia of big sun gear, J_{a1}, J_{a2}, and J_{a3} is the moment of inertia for short planet gears, and $J_{a1} = J_{a2} = J_{a3} \cdot J_{b1}$, J_{b2}, and J_{b3} is the moment of inertia for long planetary gears, and $J_{b1} = J_{b2} = J_{b3} \cdot J_r$ is the moment of inertia for ring gear.

5.2   THE TORSIONAL DYNAMIC MODEL OF THE POWER-SPLIT HYBRID SYSTEM

The TV model for the power-split hybrid driveline system is presented in Fig. 5.2.

And the engine dynamics equation can be described by

$$J_e \ddot{\theta}_e + k_{tc} (\theta_e - \theta_c) = 0, \tag{5.52}$$

where J_e is the moment of inertia of the engine, θ_e is the angular displacement of engine, and θ_c is the angular displacement of planetary carrier. $k_{tc} = \dfrac{k_t \cdot k_c}{k_t + k_c}$, k_t, and k_c are the CPGS of the torsional damper and the planetary carrier. The dynamics equation of the TV can be described by

$$J_m \ddot{\theta}_m + k_{rm}r_{m1}(\theta_m r_{m1} - \theta_r r_{r2}) + k_{md}r_{r2}(\theta_m r_{m2} - \theta_d r_d) = 0, \tag{5.53}$$

where J_m is the moment of inertia of the reducer, k_{rm} is meshing stiffness between the ring and the reducer, k_{md} is meshing stiffness between the reducer and the differential. r_{m1} and r_{m2} are base radiuses of reducer gears, r_{r2} is the base radius of the outer ring, and θ_m and θ_r are angular displacements of the reducer and ring, respectively.

The equilibrium equation of TV for the differential can be derived as follows:

$$J_d \ddot{\theta}_d + k_{md}r_d (\theta_d r_d - \theta_m r_{m2}) + k_{lh}(\theta_d - \theta_{lw}) + k_{rh}(\theta_d - \theta_{rw}) = 0, \tag{5.54}$$

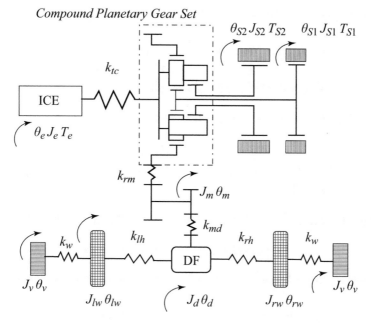

Figure 5.2: TV model of hybrid driveline [49].

where J_d is moment of inertia of the differential, k_{lh} and k_{rh} are TS of the left and right half shafts, θ_d, θ_{lw}, and θ_{rw} are angular displacements of the differential, left and right wheels, respectively.

The equilibrium equation of TV for left wheel may be given as follows:

$$J_{lw}\ddot{\theta}_{lw} + k_{lh}(\theta_{lw} - \theta_d) + k_w(\theta_{lw} - \theta_v) = 0, \qquad (5.55)$$

where J_{lw} is the moment of inertia of the left wheel, k_w is TS of the wheel, θ_v is angular displacement of the vehicle.

The equilibrium equation of TV for the right wheel for is written as follows:

$$J_{rw}\ddot{\theta}_{rw} + k_{rh}(\theta_{rw} - \theta_d) + k_w(\theta_{rw} - \theta_v) = 0, \qquad (5.56)$$

where J_{rw} is the moment of inertia of the right wheel.

The equilibrium equation of TV for the vehicle can be written as

$$J_v\ddot{\theta}_v + k_w(\theta_v - \theta_{lw}) + k_w(\theta_v - \theta_{rw}) = 0, \qquad (5.57)$$

where $J_v = m_v \cdot r_w^2$, m_v is the mass of the vehicle, r_w is the radius of the wheel.

By combining the equations for the compound planetary gear set in Eq. (5.9) with the equations of motion for the driveline components in Eqs. (5.52)–(5.57), the equations of motion for the power-split hybrid system can be written in matrix form as

$$M\ddot{x} + Kx = 0, \tag{5.58}$$

where M and K are mass and stiffness matrices of 16 order, and $\{x\}$ is 16 order vector of generalized displacement, and presented for details as follows:

$$X = [\; \theta_e \quad \theta_c \quad \theta_{S1} \quad \theta_{S2} \quad \theta_r \quad \theta_{a1} \quad \theta_{a2} \quad \theta_{a3} \quad \theta_{b1} \quad \theta_{b2} \quad \theta_{b3} \quad \theta_m \quad \theta_d \quad \theta_{lw} \quad \theta_{rw} \quad \theta_v \;]^T. \tag{5.59}$$

5.3 NUMERICAL ANALYSIS OF NATURAL FREQUENCIES AND MODES

A harmonic solution to Eq. (5.58) is supposed to be the following form as:

$$\{x\} = \sin \omega t \, \{u\}, \qquad \{\ddot{x}\} = \omega^2 \sin \omega t \, \{u\}. \tag{5.60}$$

By replacing Eq. (5.60) into Eq. (5.58), the following standard eigenvalue equation can be obtained:

$$[K]_i \, \{u\}_r = \omega_r^2 \, [M]_i \, \{u\}_r, \quad r = 1, 2 \ldots N. \tag{5.61}$$

With Eq. (5.61), numerical simulation results for natural frequencies and eigenvectors of the power-split hybrid driveline system are obtained.

The frequencies corresponding to the pure electric drive mode and the hybrid drive mode are presented in Tables 5.1 and 5.2.

Table 5.1: Natural frequencies of the hybrid system in the pure electric driving condition

Mode order	1	2	3	4	5	6
Frequency (Hz)	5.37	26.12	26.64	410.3	1584.0	2890.0
Mode order	7	8	9	10	11	12
Frequency (Hz)	4264.0	10023.6	10023.6	10168.7	16461.60	16461.60
Mode order	13					
Frequency (Hz)	16566.20					

Figures 5.3–5.6 describe the hybrid driveline mode shapes in pure electric driving mode, and Figs. 6.7–6.11 depict the mode shapes in hybrid driving mode. Where $e, S1, S2, a1, a2, a3, b1, b2, b3, r, red, diff, lw, rw$, and v denote the engine, sun gear 1, sun gear 2, three short planets, three long planets, ring, reducer, differential, left wheel, right wheel and vehicle, respectively [36].

Figure 5.3 shows that the first-order eigen mode is the rigid mode, the second order is related to the TV of driving wheels with respect to the half shafts, the third order is relevant

Table 5.2: Natural frequencies of the hybrid system in the hybrid driving condition

Mode order	1	2	3	4	5	6
Frequency (Hz)	5.6	17.3	26.1	27.1	836.1	2076.0
Mode order	7	8	9	10	11	12
Frequency (Hz)	3722.0	4441.0	10023.6	10023.6	10508.6	16461.5
Mode order	13	14				
Frequency (Hz)	16461.5	16646.5				

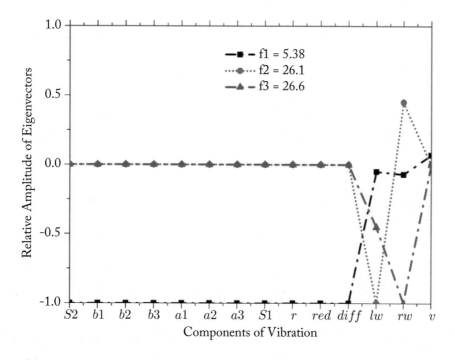

Figure 5.3: The first, second, and third eigen modes of TVs of the drivetrain in the pure electric driving condition.

to the TV of the wheels in opposite directions, and the fourth mode represents the TV of the wheels in the same direction. As shown in Fig. 5.4, the fifth-order refers to the sun gears, the sixth order is the vibration of the differential, the seventh order is a coupled TV of the ring, the reducer gear and the differential, and the vibration of the reducer gear plays a leading role in the eighth order mode. As shown in Figs. 5.5 and 5.6, the ninth and tenth orders are corresponding to the TVs of the long planets, the eleventh order reflects the TV of the long and short planets

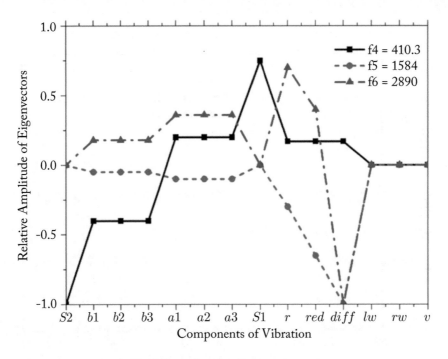

Figure 5.4: The fourth through sixth eigen modes of TVs of the drivetrain in pure electric driving condition.

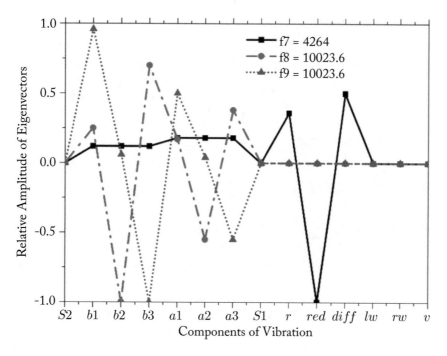

Figure 5.5: The seventh through ninth eigen modes of TVs of the drivetrain in the pure electric driving condition.

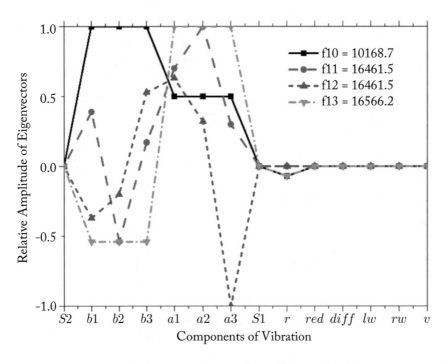

Figure 5.6: The tenth through thirteenth eigen modes of TVs of drivetrain in the pure electric driving condition.

in the same direction, the twelfth and thirteenth orders are the TV of the short planets, and the fourteenth order is the TV of the long and short planets in opposite directions.

It can be seen from Figs. 5.3–5.6 that the amplitude of the whole drive train is larger that of in the first-order mode, the left-wheel amplitude is the largest in the second-order mode, and the right wheel also has a large amplitude; the third-order mode has the largest amplitude and the left wheel amplitude is larger; the fourth-order mode has the largest amplitude of the large sun gear, and the amplitude of the small sun gear is obvious; the fifth-order mode corresponds to the differential. The reducer also has large amplitude; the sixth-order mode corresponds to the differential, but the ring gear and the reducer also have a large amplitude. The amplitude of the reducer is the largest at the seventh-order natural frequency, and the differential and the ring gear also have a large amplitude; the amplitude of the long planetary gear 2 is larger at the eighth-order natural frequency, and the long planetary gear 3 has a larger reverse vibration. The three long planetary gears at the tenth order mode has the largest amplitude, and the three short planetary wheels also have a large amplitude, the eleventh natural frequency has the largest amplitude of the short planetary gear 2, and the short planetary gear 3 also has a large amplitude. At the twelfth-order mode, the short planetary gear 3 has the largest amplitude, and the short planetary gear 1 and the long planetary gear 3 also have a large amplitude; the three short planetary gears have the largest amplitude at the thirteenth natural frequency, and the three long planetary gears also have large amplitude. Comparing with the experimental results, it can be found that the fifth and sixth orders in pure electric working conditions are the important reasons for the vibration and noise of the drive train. Figures 5.7–5.11 show the mode shapes of the powertrain in hybrid driving mode.

As can be seen from Fig. 5.7, in the hybrid driving mode, the low orders of eigen modes and frequencies are also referred to the vehicle and driving wheels. As shown from Fig. 5.8, the sixth mode seems to be a coupled vibration corresponding to differential, carrier, reducer, and ring, the seventh mode corresponds TV of the differential, the eighth mode represents TV of the short and long planets, the ninth mode is referred to the TV of the long and short planets and the differential, and the tenth to fifteenth modes are corresponding to higher frequency TVs of short planets and long planets.

It can be concluded that the low-order frequencies are mainly relevant to the vehicle and wheels, the middle frequencies are related to the TV of sun gears, differential and reducer, and the high orders are concentrated on differential, reducer, and planets in both driving mode.

It can be seen from Figs. 5.7–5.11 that the amplitude of the entire drive train is larger in the first mode, the amplitude of the engine and the left wheel is the largest at the second-order natural frequency, the left and right wheels have the largest amplitude at the third-order natural frequency. In the fourth-order mode, the amplitude of the planet carrier, the large sun gear, the three long planetary gears, the three short planetary gears, the ring gear, the speed reducer, and the differential are relatively large. In the fifth-order mode, the amplitude of the differential is the largest, the amplitude of the carrier, the reducer, and the ring gear is large,

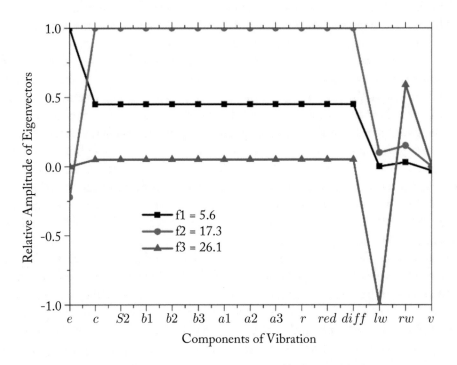

Figure 5.7: The first, second, and third eigen modes of TVs of drivetrain in the hybrid driving condition.

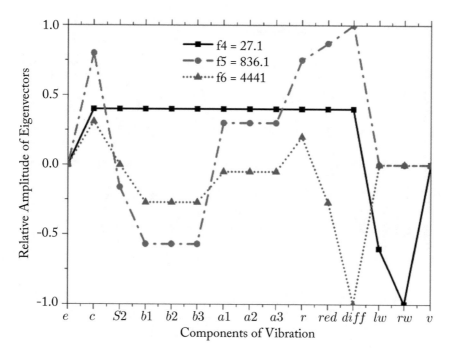

Figure 5.8: The fourth through sixth eigen modes of TVs of the drivetrain in the hybrid driving condition.

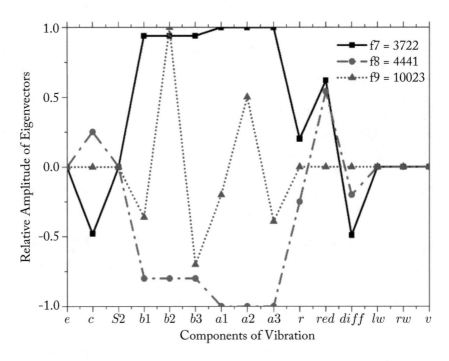

Figure 5.9: The seventh through ninth eigen modes of TVs of the drivetrain in the hybrid driving condition.

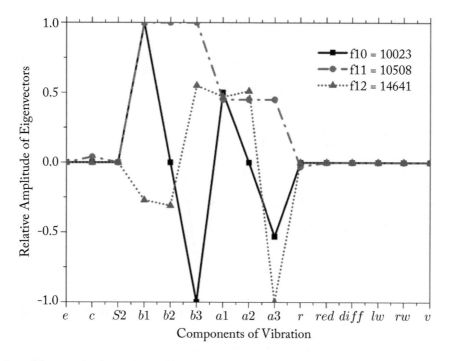

Figure 5.10: The tenth through twelfth eigen modes of TVs of the drivetrain in the hybrid driving condition.

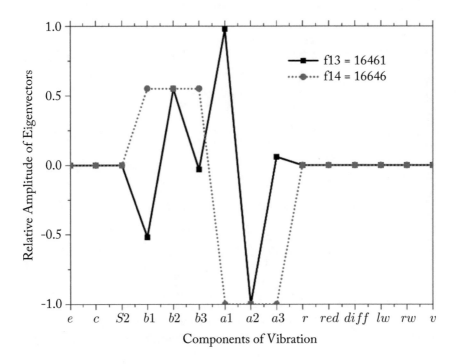

Figure 5.11: The thirteenth through fourteenth eigen modes of TVs of the drivetrain in the hybrid driving condition.

and the amplitude of the differential is the largest in the sixth-order mode. At the seventh-order natural frequency, the amplitudes of the three long planetary gears and the three short planetary gears are the largest, and the amplitudes of the differential and the reducer are relatively large. In the eighth-order mode, the amplitude of the three short planetary gears is the largest, and the amplitudes of the three long planetary gears are larger, the amplitude of the long planetary gears 2 is the largest in the ninth-order mode, and the amplitude of the long planetary gears 3 and the short planetary gears 2 is large. In the tenth-order mode, the long planetary wheel 1 and the long planetary gear 3 have the largest amplitude; in the eleventh mode, the long planetary gear has the largest amplitude, and the short planetary gear has a large amplitude. At the twelfth order natural frequency, the short planetary gear 3 has the largest amplitude, and the short planetary wheel 1, the short planetary gear 2 and the three long planetary wheels have a large amplitude. In the thirteenth mode, the amplitudes of the short planetary gear 1 and the short planetary gear 2 are the largest, and the amplitudes of the long planetary gear 1 and the long planetary gear 2 are large; the amplitude of the three short planetary gears at the natural frequency of the fourteenth order is the largest. The amplitude of the three long planetary wheels is large.

5.4   CONCLUSION

The TV dynamics model of the hybrid powertrain was established. The dynamic equations of the various components of the drive train were established. The natural frequencies and their vibration modes of the hybrid powertrain in pure electric and hybrid modes were calculated. The results showed the following.

In the pure electric working condition, the amplitude of the whole vehicle, the left wheel and the right wheel was relatively large at the low-order natural frequency, and the TV of the large and small sun gear, the differential and the reducer gear was sharp at the intermediate frequency mode. The long and short planetary gears also had smaller TVs. The TVs of the long and short planetary gears were mainly reflected at the high-order modes. In the hybrid power condition, the vibrations of the left and right wheels, the whole vehicle and the engine were relatively large at low natural frequencies. The TVs of large and small sun gears, differentials and reducers, and the engine occured at the intermediate frequency mode mainly. The TVs of the long and short planet wheels were mainly reflected in the high-order modes. Comparing with the experimental results, it was found that the fifth and sixth orders that in pure electric working conditions were the important reasons for the vibration and noise of the drive train.

CHAPTER 6

Modeling of the Hybrid Powertrain with ADAMS

In Chapter 5, the hybrid powertrain was mathematically modeled and modal calculations were performed using MATLAB software. In this chapter, the hybrid powertrain was modeled by ADAMS software, and the results of modal calculations and mathematical modeling were performed well. In contrast, the calculation results are in good agreement, which verifies the correctness of the model. On the basis of verifying the correct model, the forced TV simulation analysis of the hybrid powertrain is carried out [81]. By applying the excitation to the engine and the motor and changing the parameters of the powertrain components, the transfer function method is used to find the main elements on the TV, which provide a reference for improving the TV of hybrid vehicles. Meanwhile, so as to reduce the TV of the hybrid powertrain, add the DMF to the powertrain, and the parameter combination that minimizes the TV of the powertrain is obtained.

6.1 MODELING OF THE HYBRID POWERTRAIN WITH ADAMS

In the ADAMS, the hybrid system dynamics model is established, and the constrained Lagrangian dynamics method is adopted. We present several hypotheses for building the dynamics model to assess the TV of the hybrid drivetrain using ADAMS software [81].

1. The axle's moment of inertia (MOI) equals to two connected lumped masses through the axle.

2. The crankshaft's MOI and rotating components in the engine are calculated to the flywheel's rotational inertia.

3. Neglect the effects of the TV caused by longitudinal, lateral, and vertical vibrations.

4. The torsional damper is used as a torsional spring.

5. Ignore errors of gear cutting and installation and also the wear deformation in powertrain systems.

6. Ignore the impact of accessory equipment.

7. Ignore the deformation of the gear.

In particular, the key to building the powertrain model is to build a dynamic model of the CPGS. In ADAMS, conventional gear joints cannot calculate the TV characteristics of the gear train. However, this work uses the virtual gear pair to conduct research. An assist massless gear is added to connect the driving gear, and the driven gear is connected to the assist gear. Power is transferred from the driving gear to the virtual gear and then transferred to the driven gear. The virtual gear pair's working principle is demonstrated below. Figure 6.1 indicates the transition from the meshing to the torsional stiffness (TS) of the two gears [81]. As shown in Fig. 6.1, k_m is the two gears' meshing stiffness (MS) and k_t is the torsional spring's stiffness, r is the radius of the driven gear's pitch circle. The driven gear loaded torque is expressed as:

$$T = k_m \cdot \Delta r \cdot r \tag{6.1}$$

$$T = k_t \cdot \Delta\theta, \tag{6.2}$$

where $\Delta\theta = \frac{\Delta r}{r}$. Consequently, the two gears' M.S. will be converted into the spring's TS and $k_t = k_m \cdot r^2$. Power is transferred from the driving gear to the virtual gear through the conventional gear pair and then transmitted to the driven gear through the torsional spring in virtual gear pair. Here, the M.S. is the synthesizing stiffness calculated by ishikawa formula and the M.S. is considered as a constant [37].

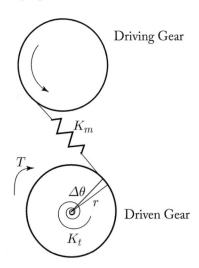

Figure 6.1: The transformation of MS [81].

The compound planetary gear rows for the power-split device consist of two planetary. The first row consists of the sun gear 1, short and planet ring, and the second row consists of the long planet, the sun gear 2, short ring and planet. Thus, in the CPGS's dynamic model, the virtual gear is connected to the driven gear with the torsional of the long planet, and the sun gear $S2$ meshes with the long planet's virtual gear. The stiffness of the torsional spring between the long planet and sun gear 2 can be demonstrated as below:

$$k_{t-S2pl} = k_{S2pl} \cdot r_{long}^2, \tag{6.3}$$

where r_{long} is the radius of the long planet's pitch circle the and k_{S2pl} is the M.S. between the long planet and sun gear 2.

As the driven gear, the short planets mesh with the long planet and sun gear 1. Consequently, the torsional spring's stiffness of can be expressed as:

$$k_{t-S1ps} = k_{S1ps} r_{short}^2 \tag{6.4}$$
$$k_{t-pspl} = k_{m3} r_{short}^2, \tag{6.5}$$

where k_{pspl} is the M.S. between the long and short planet, the k_{S1ps} is M.S. between the sun gear 1, and short planet, r_{short} is the radius of the long planet's pitch circle.

In the same way, building three virtual gear of the ring to mesh with the short planets, respectively, and the virtual gears are connected to the ring by the torsional spring, respectively. Consequently, the stiffness of the torsional spring between short planet and ring can be described as:

$$k_{t-rps} = k_{rps} r_{ring}^2, \tag{6.6}$$

where r_{ring} is the radius of ring gear's the pitch circle and k_{rps} is M.S. between the short planet and ring.

The power-split HEV powertrain's multi-body dynamic model (MBDM) built in ADAMS is shown in Fig. 6.2. The model consists of the engine, CPGS, torsional damper, differential, reducer, wheels, half shafts, and vehicle. The CPGS is made up of a ring, a carrier, a small sun gear, three short and long planets, and a big sun gear. The MOI of $E2$ and $E1$ are loaded on the big sun and small sun gear separately. The carrier and flywheel are connected by the torsional damper as shown in Fig. 6.2, which is represented by a torsional spring that has torsional damping and stiffness. Right and left wheels are connected to the differential through the right and left half shaft separately, and the stiffness of the left and right half shafts, respectively, 5520 and 4420 N·m/rad separately. The wheels are connected to the equivalent MOI of the vehicle by torsional springs (stiffness is 44692 N·m/rad).

The following equation demonstrates the equivalent MOI J_{car}:

$$J_{car} = m_{car} \cdot r_{wheel}^2, \tag{6.7}$$

where r_{wheel} is the wheel radius and m_{car} is the vehicle mass.

The natural frequencies are computed in the vibration module in the software ADAMS. After establishing the powertrain's dynamic model, by inserting the vibration module into the main module, the eigenvectors and frequencies can be obtained. In the Post Processor module, the pattern shape can be shown.

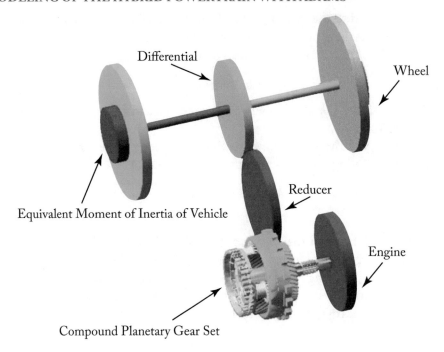

Figure 6.2: Hybrid powertrain torsional dynamic model.

6.2   COMPARISON AND VERIFICATION OF THE TWO MODELS

In MATLAB, there are differential solvers for equations. Ode45 is the preferred choice for calculating equations according to MATLAB recommendations. The ode45 algorithm is a single step method which is based on the explicit Runge–Kutta formula. The ode15s algorithm is a multi-step method which is a variable solver.

ADAMS also has various kinds of solvers, such as Dstiff, Gstiff, Wstiff, Constant-BDF, RKF45, and ABAM. The advantages of Gstiff solver are high precision and less time-consuming. This algorithm could be applied to many simulation issues.

Chapters 4 and 5 give the detailed model process of the hybrid powertrain with MATLAB an ADAMS, respectively. Tables 6.1 and 6.2 demonstrate the natural frequencies computed from hybrid and pure electric driving mode, separately [36].

Table 6.1 demonstrates that the calculation of the hybrid powertrain's natural frequency in pure electric mode by ADAMS and MATLAB shows that the first-order to fourth-order's low-order natural frequency calculation error is small; fifth-, sixth-, and seventh-order natural frequency calculation results are relatively large, the fifth order is 4.5%, the sixth order is 3.9%, and the seventh order is 5.4%, but these errors are within the engineering allowable range, and

Table 6.1: Hybrid powertrain's natural frequency in the pure electric driving mode

Frequency (Hz)	MATLAB	ADAMS
1st	5.378	5.393
2nd	26.12	26.12
3rd	26.64	26.65
4th	410.3	404.2
5th	1584.0	1513.75
6th	2890.0	2776.9
7th	4264.0	4036.2
8th	10023.6	9844.6
9th	10023.6	9844.6
10th	10168.7	9989.0
11th	16461.5	16290.9
12th	16461.5	16290.9
13th	16566.2	16398.6

the error of calculation result at thigh-order natural frequency is smaller, all within 2%. From the analysis of the settlement results, it can be seen that in the pure electric mode, the results calculated by the two methods mutually verify the correctness of the model.

It can be seen from Table 6.2 that the calculation of the natural frequency of the hybrid powertrain in the hybrid mode by MATLAB and ADAMS shows that the first-order to fourth-order's low-order natural frequency calculation errors are small, first-order, third-order, and fourth-order natural frequency calculation has no error; the sixth-order, seventh-order and eighth-order natural frequency calculation results have relatively large errors, the sixth order is 5.9%, the sixth order is 4.6%, and the seventh order is 3.5%. However, these errors are within the engineering allowable range, and the high-order natural frequency calculation results have small errors, all within 2%. From the analysis of the settlement results, it can be seen that in the hybrid mode, the results of the two methods also verify the correctness of the model.

As shown in Tables 6.1 and 6.2, two kinds of approaches are adopted to explore the noise characteristics and TV of a power-split hybrid powertrain. The first approach is to establish the MBDM in ADAMS. The second approach is to develop the mathematical model of the powertrain with consideration of equilibrium equations of TV. The vibration modes and natural frequencies computed from MATLAB and ADAMS agree well, demonstrating that two approaches are both accurate in describing the torsional characteristics.

Table 6.2: Hybrid powertrain's natural frequency in the hybrid driving mode

Frequency (Hz)	MATLAB	ADAMS
1st	5.6	5.6
2nd	17.3	17.4
3rd	26.1	26.1
4th	27.1	27.1
5th	836.1	823.3
6th	2076.0	1951.5
7th	3722.0	3550.6
8th	4441.0	4286.8
9th	10023.6	9844.6
10th	10023.6	9844.6
11th	10508.6	10314.1
12th	16461.5	16290.9
13th	16461.5	16290.9
14th	16646.5	16470.6

6.3   ANALYSIS ON THE FORCED VIBRATION (FV)

Forced Vibration (FV) is described as the system vibration under the periodic force. FV not only relates to the frequency and amplitude of the driving force, but also the vibration system's natural frequency. When the driving frequency equals to the natural frequency, the resonance happens.

By comparison between the excitation in the FV and the response of TV, the TV is promised to be minimum. Therefore, the NVH quality and ride property has been increased.

This section presents the analysis of the factors that influence the TV by adopting the FV simulation. The angular acceleration is taken as an output as it is important to evaluate TV. In the time domain, the input excitation can be demonstrated as below [43]:

$$f(t) = F_0 \left[\cos(\omega \cdot t + \theta) + j \cdot \sin(\omega \cdot t + \theta) \right], \tag{6.8}$$

where $F_0 = 100$ N·m is the amplitude of input, θ is the phase angle ($\theta = 0$), $\omega = 0 \sim 100$ Hz is the frequency range of input, and t is the time.

Angular acceleration is an important index to assess the TV. Consequently, we use an angular acceleration response of transmission components as the output to carry out the FV analysis.

6.3.1 INFLUENCE OF VARYING DAMPING OF TORSIONAL DAMPER ON FREQUENCY RESPONSE

With the aim to analyze damping of torsional damper that influences the powertrain torsional vibration (TV) under different kinds of excitations, the damping of the torsional damper is set to 0.5, 0.75, 1, 1.25, and 1.50 of related parameters in the simulations. The torsional damper' equivalent damping is 10 N·m·s/rad. Consequently, the damping is set to 5, 7.5, 10, 12.5, and 15 N·m·s/rad.

1. The engine is taken as input excitation.

   When the engine provides the driving force and changes the TV damper damping value for FV analysis, the TV frequency domain response of the planet carrier changes and the forced analysis results are shown in Fig. 6.3.

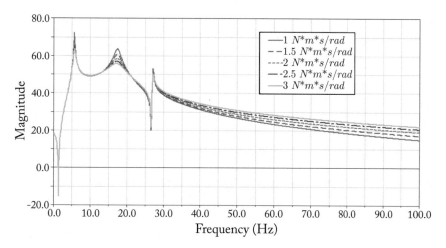

Figure 6.3: Influence of damping on frequency response under engine excitation.

It can be seen from Fig. 6.3 that when the drive train is excited by the engine, the change of the TV damper damping will cause a change in the resonance peak. The larger the damping, the smaller the resonance peak of the drive train and more obvious the first and second resonance peak changes relatively.

When the engine provides the driving force, the differential frequency response of the torsion damper is shown in Fig. 6.4.

It can be seen from Fig. 6.4 that when the drive train is under the excitation of the engine, the change of the TV damper damping will cause the change of the resonance peak of the differential frequency domain response. The larger the damping, the smaller the resonance peak of the drive train, the first and the second resonance peak of which change is comparatively obvious. It is also shown in Fig. 6.4 that the change in the damping value

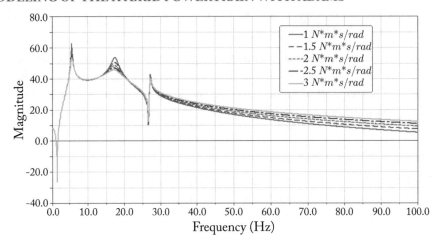

Figure 6.4: Influence of damping on frequency response under the engine excitation.

of the torsional damper does not cause a change in the natural frequency of the hybrid powertrain.

2. The electric motor is chosen as input excitation.

In this part, the large motor is used as the excitation source to calculate the FV, and the frequency response of the torsional damper to the planetary frame and differential in the transmission system is analyzed. Furthermore, find the law of the influence of the TV of the torsion damper on the TV of the drive train under the excitation of a large motor.

Figure 6.5 shows the effect of torsional damper damping characteristics on the frequency domain response of the planetary frame excited by a large motor. The effect of torsional damper damping characteristics on the frequency domain response of the differential excited by a large motor is shown in Fig. 6.6.

From the results of Figs. 6.5 and 6.6, it can be seen that the response peak value of transmission system components at resonance point decreases with the increase of torsional damper damping under the excitation of a large motor, and the peak value of the second resonance point decreases obviously. It is also shown from the graph that the damping of torsional damper has no effect on the response of high-frequency stage. In addition, the damping of torsion damper does not affect the natural frequency of the hybrid power train.

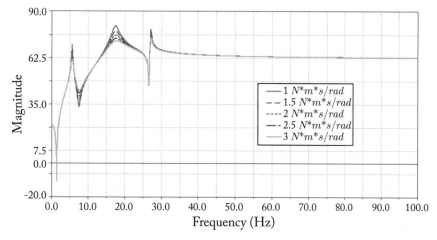

Figure 6.5: Influence of damping on frequency response under the excitation of motor 2.

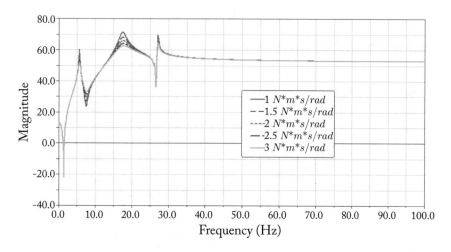

Figure 6.6: Influence of damping on frequency response under the excitation of motor 2.

6.3.2 INFLUENCE OF VARYING STIFFNESS OF TORSIONAL DAMPER ON FREQUENCY RESPONSE

With the aim to analyze the damper's TS that influences the powertrain TV, the TS of torsional damper is set to 0.5, 0.75, 1, 1.25, and 1.50 of related parameters. The TS of the torsional damper is 618. Consequently, the TS of torsional damper is set to 309, 463, 618, 772, and 972 N·m·s/rad.

1. The engine is taken as input excitation.

The engine is used as the excitation source, and the TS of the TV damper is changed to perform the FV analysis. The frequency domain response of the planet carrier is shown in Fig. 6.7, and the frequency domain response of the differential is shown in Fig. 6.8.

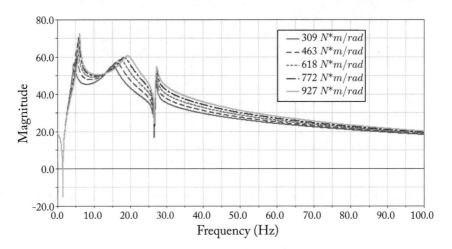

Figure 6.7: Influence of stiffness on frequency response under the engine excitation.

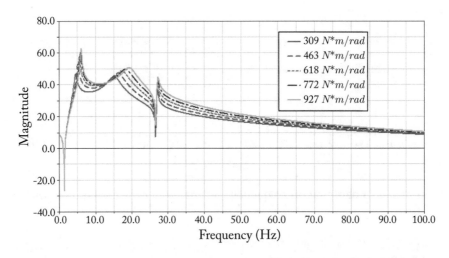

Figure 6.8: Influence of stiffness on frequency response under the engine excitation.

Figures 6.7 and 6.8 demonstrate that increasing the stiffness of the TV damper makes the frequency domain response amplitude of the powertrain larger, and reducing the TS of the TV damper can reduce the TV of the drive train, especially in the low-frequency phase, the ride comfort of the car can be improved.

It can be seen from Figs. 6.7 and 6.8 that increasing the stiffness of the TV damper makes the frequency domain response amplitude of the powertrain larger, and reducing the TS of the TV damper can reduce the TV of the drive train, especially in the low-frequency phase, the ride comfort of the car can be improved. In addition, it can be concluded from Figs. 6.7 and 6.8 that increasing the torsional rigidity of the TV damper causes the first and second natural frequencies of the powertrain to change, and the TS is larger, the greater the natural frequency of the first order and the second order, and the change in stiffness has no effect on the third-order natural frequency.

2. The electric motor is taken as input excitation.

The large motor is used as the excitation source to change the TS of the TV damper. The FV analysis is finished in the ADAMS dynamic model. The frequency domain response of the planet carrier is shown in Fig. 6.9. The frequency domain response of the differential is shown in the figure.

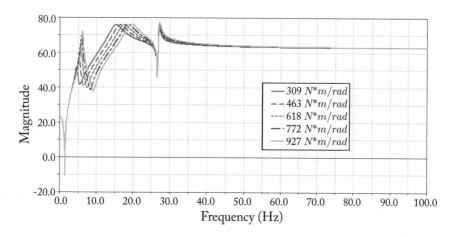

Figure 6.9: Influence of stiffness on frequency response under the excitation of motor 2.

It can be seen from Figs. 6.9 and 6.10 that under the excitation of the large sun gear, the change of the TV of the torsion damper causes the first and second natural frequencies of the powertrain to change, and the greater the stiffness, the greater the natural frequency; the conclusion can be drawn that increasing the stiffness of the TV damper makes the first-order resonance peak become larger, and the second-order peak does not change; under the excitation of the large motor, increasing the TS of the TV damper to the high-frequency response of the transmission system has no effect.

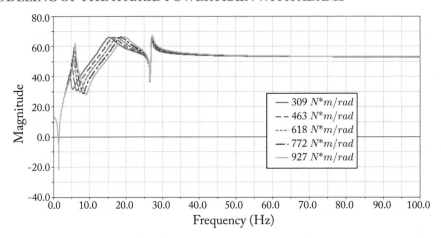

Figure 6.10: Influence of stiffness on frequency response under the excitation of motor 2.

6.3.3 INFLUENCE OF THE FLYWHEEL'S MOI ON FREQUENCY RESPONSE

With the aim to analyze the MOI of flywheel that influences the powertrain TV under different excitations, the MOI of flywheel is set to 0.5, 0.75, 1, 1.25, and 1.50 of related parameters in the simulations. The MOI of the flywheel is 0.28 kg·m². Consequently, the MOI of flywheel is set to 0.14, 0.21, 0.28, 0.35, and 0.42 kg·m².

Using the engine as the excitation source, the flywheel inertia is changed in the ADAMS dynamics model for FV analysis. The frequency domain response of the planet carrier is shown in Fig. 6.11. The frequency domain response of the differential is shown in Fig. 6.12.

From Figs. 6.11 and 6.12, we can see that the change of the MOI of the flywheel causes the first order natural frequency to change. The larger the flywheel's MOI, the smaller the first order natural frequency of the powertrain. In addition, it is also found that the when the engine becomes an input excitation, increasing the flywheel's MOI of can reduce the response amplitude of the carrier and the differential over the entire frequency band. Consequently, for the hybrid powertrain, increasing the flywheel's MOI is beneficial to reduce the twisting vibration of the drive train, which improves the ride comfort performance of the car.

6.3.4 INFLUENCE OF VARYING STIFFNESS OF HALF SHAFT ON FREQUENCY RESPONSE

With the aim to analyze TS of half shaft that influence the powertrain TV, the TS of half shaft are set to 0.5, 0.75, 1, 1.25, and 1.50 of related parameters in the simulations. The TS the half shaft are 5520 and 4222 N·m/rad, respectively. Consequently, the stiffness of the left half shaft

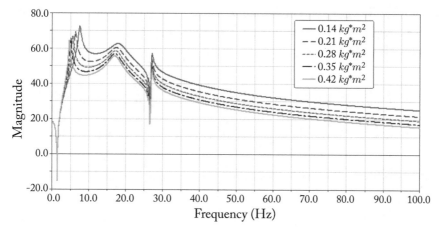

Figure 6.11: Influence of rotational inertia on frequency response under the engine excitation.

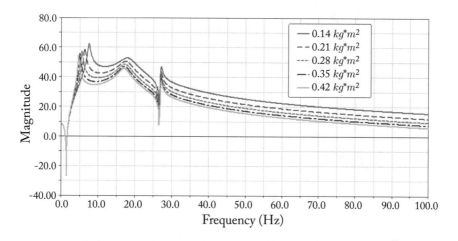

Figure 6.12: Influence of rotational inertia on frequency response under the engine excitation.

is preset to 2760, 4140, 5520, 6900, and 8280, the right half shaft is supposed to 2760, 4140, 5520, 6900, and 8280 N·m/rad.

The engine is used as the excitation source, and the right half-axis TS is changed in the ADAMS dynamic model for FV analysis. The frequency domain response of the differential is shown in Fig. 6.13. The large motor is used as the excitation source in the ADAMS dynamics model, the TS of the right half-shaft is changed to perform FV analysis. The frequency domain response of the differential is shown in Fig. 6.14.

The engine is used as the excitation source, and the left half-axis TS is changed in the ADAMS dynamic model for FV analysis. The frequency domain response of the differential is shown in Fig. 6.15. The large motor is used as the excitation source in the ADAMS dynamics model. The FV analysis of the left half-axis TS is performed, and the frequency domain response of the differential is shown in Fig. 6.16.

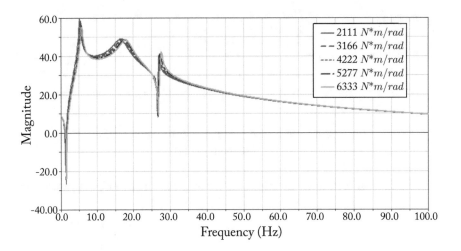

Figure 6.13: Influence of right half shaft stiffness on the frequency response of the differential TVs under the engine excitation.

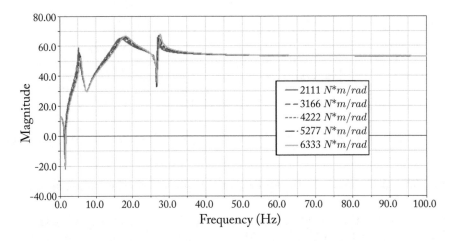

Figure 6.14: Influence of right half shaft stiffness on the frequency response of the differential TVs under the excitation of motor 2.

It can be seen from Figs. 6.13–6.16 that changing the TS of the half shaft can cause a small change in the natural frequency of the powertrain, and the change in the stiffness of the

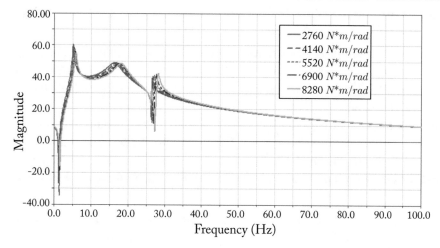

Figure 6.15: Influence of left half shaft stiffness on the frequency response of the differential TVs under the engine excitation.

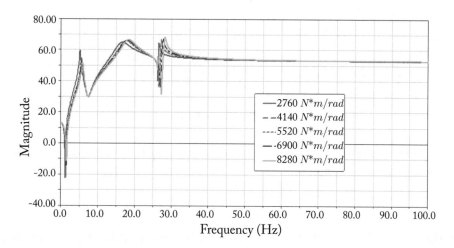

Figure 6.16: Influence of left half shaft stiffness on the frequency response of the differential TVs under the excitation of motor 2.

half shaft has a small change in the response amplitude of the low frequency of the powertrain, while it does not affect the response amplitude of the frequency band after 35 Hz.

6.3.5   INFLUENCE OF VARYING DAMPING OF HALF SHAFT ON FREQUENCY RESPONSE

With the aim to analyze the torsional damping of half shaft that influences the powertrain TV, the torsional damping of half shaft is set to 0.5, 0.75, 1, 1.25, and 1.50 of related parameters. The TS the half shaft is 1 N·m·s/rad. Consequently, the damping of the left half shaft is supposed to 0.5, 1, and 2 N·m·s/rad.

Using the engine as the excitation source, the left half-axis damping is changed in the ADAMS dynamics model to perform the FV analysis, and the frequency domain response of the differential is shown in Fig. 6.17. The right half-axis TS is changed to perform the FV analysis and the frequency domain response of the differential is shown in Fig. 6.18.

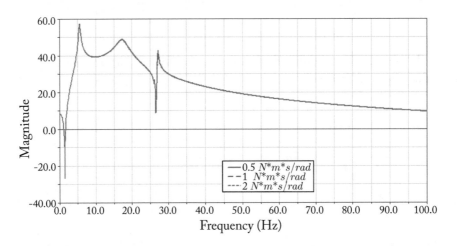

Figure 6.17: Influence of left half shaft damping on the frequency response of the differential TVs under the excitation of engine.

It can be seen from Figs. 6.17 and 6.18 that changing the half shaft damping has little effect on the powertrain. Regardless of the resonance frequency or the response amplitude, the left and right half shaft damping variations hardly change the response of each frequency band of the hybrid powertrain.

6.3.6   INFLUENCE OF WHEEL TS VARIATION ON THE TV OF POWERTRAIN

In order to analyze the factors that influence the powertrain TV, the TS of left half shaft is set to 0.5, 1, and 2 of related parameters in the simulations. The TS of the half shaft is 44,690 N·m/rad. Consequently, the stiffness of the left half shaft is set to 22,345, 44,690, and 89,380 N·m/rad.

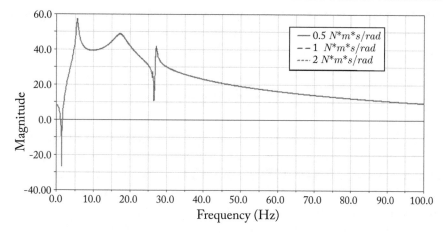

Figure 6.18: Influence of right half shaft damping on the frequency response of the differential TVs under the excitation of engine.

Using the engine as the excitation source, the TS of the wheel is changed in the ADAMS dynamics model for FV analysis. The frequency domain response of the differential is shown in Fig. 6.19.

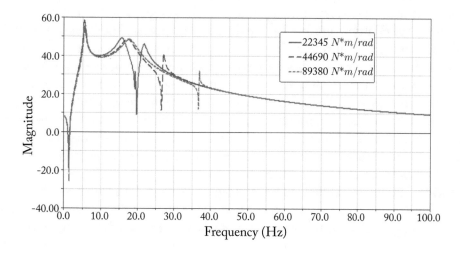

Figure 6.19: Influence of tire stiffness on frequency response of the differential TVs under the excitation of engine.

It can be seen from Fig. 6.19 that changing the TS of the wheel causes the natural frequencies of the second and third orders to change, mainly to change the natural frequency of the third order; increasing the torsional rigidity of the wheel makes the third-order resonant frequency becomes larger, and the response amplitude becomes smaller; the change of the wheel TS has no effect on the first-order resonance frequency and the frequency band after 40 Hz.

6.4   TV CHARACTERISTICS AND OPTIMIZATION ANALYSIS OF DUAL MASS FLYWHEEL

To reduce the TV of the powertrain, this section applies the DMF TV damper to the hybrid powertrain, and compares and analyzes the TV characteristics of the clutch driven disc TV damper Clutch Torsional Damper (CTD) and the Dual Mass Flywheel (DMF) in a hybrid vehicle, the results show that the DMF damper is more conducive to reducing the TV of the hybrid powertrain. Two flywheel inertia ratio in the DMF damper and the influence of parameters such as ratio, TS, and damping on the hybrid powertrain are analyzed to obtain the optimal value of the parameters that minimize the TV of the hybrid driveline [82–90].

6.4.1   ADVANTAGE OF DMF

The traditional CTD was implemented to reduce the vehicle's TV from the start of the last century. However, the CTD generates some natural frequencies in the low-revolution speed range of engine. Fuel economy is usually sacrificed by the automakers to increase the idle speed of the engine to avoid resonance in the operation revolution speed of the engine. Based on the above reasons, the DMF development has attracted more attention. The schematics of the DMF and CTD are similar as depicted in Fig. 6.20, and both of them include elastic and damping components. The DMF has two flywheels is the main difference between them. The CTD's damper is installed on the clutch disc, and there is little space between the disc hub and the friction disc. Consequently, the distortion space of the elastic component is relatively small. Therefore, the TS is sufficient for transmitting the engine torque. The DMF damper is installed between the primary and the secondary flywheels, and the friction disc doesn't restraint space. Thus, the deformation space is sufficient, and their stiffness is large enough to reduce the TVs while avoiding any resonance of the powertrain at the engine's operating speed.

As shown in Fig. 6.20, the CTD TV damper is mounted on the clutch driven plate in the conventional automobile, and the space between the hub and the friction plate is small, so the deformation space of the spring element in the damper is small. The spring stiffness can be only designed to be large enough, in order to transmit the engine torque. The DMF damper is installed between the first flywheel and the second flywheel. The installation space is not restricted by the clutch friction plate, and the torsion spring has a large enough deformation, so the TS of the DMF damper can be designed to be small. The reduction of TS reduces the natural frequency of the drive train and avoids the idle speed of the engine, which avoids resonance with the drive train at lower engine speeds. Low stiffness not only facilitates low-frequency vibration isolation but also attenuates the high-frequency excitation of the engine to the drive train.

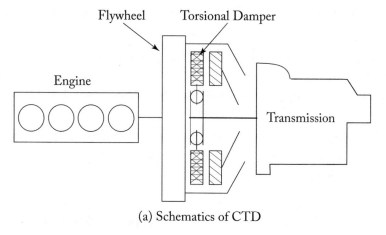

(a) Schematics of CTD

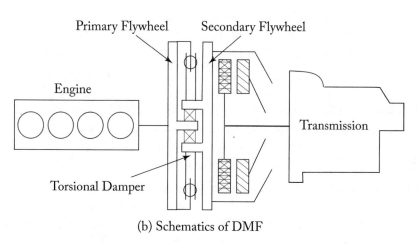

(b) Schematics of DMF

Figure 6.20: Schematics of CTD and DMF [82].

The traditional CTD damper has limited ability to solve the problem of low-frequency torsion of the car. In order to better solve the problem of ride comfort caused by TV, it is of great necessity to find a new way. In this case, the DMF damper emerges as the times require. It has been rapidly developed due to its large corner, low stiffness, and excellent vibration isolation performance. Due to the limitation of space size, the traditional CTD damper has a large TS, resulting in a large vibration transmission rate, and the vibration isolation effect is particularly poor at low frequency. The emerging DMF damper divides the flywheel into two parts, which are connected by a damper so that there is a large space for arranging the elastic elements and damping elements of the damper, enhancing the torsional rigidity of the CTD can be designed very small to reduce the TV transmitted by the engine in various frequency bands.

In comparison with conventional CTD dampers, the DMF has four greatest strengths including the following.

1. The installation space of the DMF damper is large, which overcomes the shortage of the installation space of the CTD damper so that the relative rotation angle and the transmission torque are large. The biggest feature of the DMF damper is its low TS, which greatly improves the vibration isolation performance.

2. While changing the stiffness characteristics and damping characteristics, the DMF damper can also reduce the TV of the powertrain by the MOI ratio of the first flywheel and the second flywheel.

3. The DMF application reduces the MOI of the driven disc, making shifting easier, and reducing the impact of the gear when shifting, improving the shift quality.

4. DMF damper can reduce the low-order natural frequency of the transmission, and reduce the resonance in the low engine speed, it makes it possible for the engine to work in the economic area, which is beneficial to reduce emissions and improve fuel economy.

   Disadvantages of the DMF damper include complex structure, high cost, high maintenance cost, and large axial dimension requirements for installation. However, with the development and advancement of technology, foreign DMF products have reached the requirements of the automotive industry and are widely used in vehicles.

6.4.2 DYNAMIC MODELING AND PARAMETER SELECTION OF DMF

In conventional vehicles, the DMF is the main damping component of the powertrain, which can better reduce the TV transmitted by the engine to the powertrain. This section will analyze the hybrid vehicle's TV equipped with a DMF. The research includes that analyze the influence of the MOI ratio of the first and second flywheel, damping and TS on the TV of the hybrid powertrain in the DMF, find the reason affecting the TV, and the optimal MOI ratio, TS and damping are determined to decrease the TV based on the simulation results.

Based on the dynamic model of the ADAMS hybrid powertrain illustrated in the first section of this chapter, this chapter establishes the torsional dynamics model of the ADAMS transmission with DMFs. The model created is shown in Fig. 6.21.

Compared with the first section of the hybrid powertrain model, the model established in this section is a DMF, wherein the first and second flywheel are linked by a torsion spring, and the TS is the TS of the double mass flywheel damper, damping is the damping of the damper.

There are many parameters of the hybrid powertrain. This section only analyzes the DMF characteristic parameters, including the MOI ratio of the first flywheel and the second flywheel, the TS of the DMF damper, and the damper damping. By performing FV analysis on the dynamic model, the optimal combination of parameters is selected to minimize the TV transmitted by the engine to the powertrain.

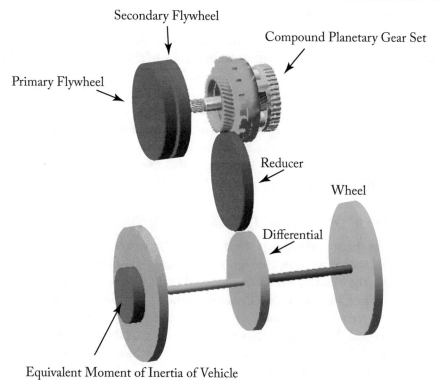

Figure 6.21: Dynamic model of a hybrid powertrain with DMF.

1. DMF MOI ratio.

   Compared with the transmission TV damper, the double mass flywheel has the most out-standing characteristic that at the MOI on the side of the damper engine is reduced, and the MOI on the side of the powertrain becomes larger. The MOI ratio of the first flywheel to the second flywheel has an optimum value such that the TV transmitted by the engine to the driveline is minimal. Determining the MOI ratio between the first and second flywheel is a key issue in the design of a DMF.

2. TS of double mass flywheel.

   The TS of the damper is one of the main parameters affecting the TV of the powertrain, which is directly associated with the ride comfort The conventional TV damper cannot reduce the torsional rigidity of the elastic member to a low level due to the space limitation, and the double mass flywheel damper has a large improvement in space, so that the elastic member can have a lower torsional rigidity. Selecting reasonable TS according to needs is beneficial to decrease the TV of the powertrain.

3. Damping of the double mass flywheel.

Friction damping is an important performance parameter of the damper, which plays a critical part in reducing the resonance peak of the TV of the powertrain. When the damper damping parameter is selected, if it is too small, the TV of the powertrain cannot be effectively reduced. The damping value is too large, which makes the damper rigidity become larger and reduces the vibration reduction performance. Consequently, when designing a DMF, analyzing the damping value to obtain the optimum is of necessity. Here, by changing the damping value of the DMF, analyzing the FV for the hybrid powertrain is attempted to find the optimal damping value so that the damper can effectively reduce the TV transmitted by the engine.

6.4.3   INFLUENCE OF THE ROTATIONAL INERTIA RATIO OF DMF

In the hybrid vehicle studied, the clutch is not used to disconnect the power, and the flywheel is directly associated with the carrier through the TV damper. Consequently, when the DMF model is established, the second flywheel is connected to the planet carrier through the original TV damper, and the first and the second flywheel are linked by a damper of the DMF.

In terms of the recent research and engineering application [89–91], the rotational inertia ratio of the primary and the secondary flywheels is between 0.024 and 2. In order to operate smoothly, the total MOI of the primary and secondary flywheels must be equivalent to that of the original flywheel. Consequently, the principal showed in Eq. (6.9) must comply when considering the moment of DMF inertia ratio:

$$
\begin{cases}
\dfrac{I_2}{I_1} \leq 2 \\[2mm]
\dfrac{I_2}{I_1} \geq 0.024 \\[2mm]
I_1 + I_2 = I_{CTD}
\end{cases}
\tag{6.9}
$$

where I_{f1} and I_{f2} indicate the primary and secondary flywheels' rotational inertias, respectively. I_{CTD} denotes the original flywheel's rotational inertia.

0.28 kg·m^2 denotes the original flywheel's rotational inertia. Different values are pre-set in the FV analysis of the powertrain to achieve the improved rotational inertia ratio. A variety of flywheel rotational inertia values studied through simulation are demonstrated in Table 6.3.

In order to obtain the optimal flywheel MOI ratio, different ratios are set in the ADAMS powertrain model for FV analysis. The values of the MOI of the first flywheel and the second flywheel are shown in Table 6.3.

Using the engine as an excitation, the first and second flywheel's MOI is changed in the dynamic model to perform FV analysis. The frequency domain response of the planet carrier is shown in Fig. 6.22, and the response of the differential is shown in Fig. 6.23. The horizontal coordinate and longitudinal coordinate indicate the frequency of the input torque loaded on the engine and the amplitudes of the frequency response, respectively

The maximum response curve in Figs. 6.22 and 6.23 is the TV response without the use of a DMF. It can be seen from the figure that the DMF can significantly reduce the TV of the powertrain. At the same time, the different moments of inertia ratio are compared and analyzed. The second and third response peaks of the powertrain components are similar when the first flywheel MOI is 0.21 kg·m² and the second flywheel MOI is 0.07 kg·m². There is no large response peak, and the overall frequency domain response curve is better. Consequently, the value of this group makes the powertrain TV optimal. The frequency domain response of the planet carrier under this set of parameters is shown in Fig. 6.24.

Table 6.3: The various values of rotational inertia of flywheels

Number of Simulation	Rotational Inertia of Primary Flywheel	Rotational Inertia of Secondary Flywheel
1	0.1 kg · m²	0.18 kg · m²
2	0.14 kg · m²	0.14 kg · m²
3	0.18 kg · m²	0.1 kg · m²
4	0.21 kg · m²	0.07 kg · m²
5	0.26 kg · m²	0.02 kg · m²

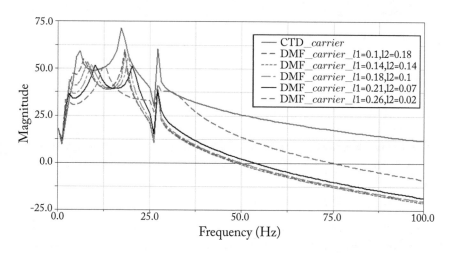

Figure 6.22: Frequency response of carrier with different rotational inertias of flywheels [43].

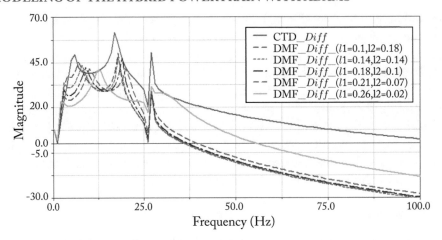

Figure 6.23: Frequency response of differential with different rotational inertias of flywheels.

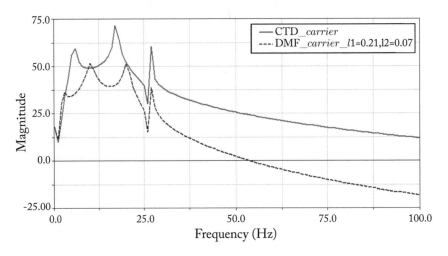

Figure 6.24: Frequency response of carrier with the best value of rotational inertias.

6.4.4   INFLUENCE OF THE TS OF DMF

Due to the available space, the length and the radius of the spring in the DMF are greater than those in CTD. Consequently, it can decrease the required stiffness of the DMF. Based on literature and experimental results [89–91], the DMF's stiffness can be denoted as follows:

$$0.1 \cdot k_{CTD} \leq k_{DMF} \leq k_{CTD}, \tag{6.10}$$

where k_{CTD} indicates the CTD's TS, and k_{DMF} demonstrates DMF's TS. The TS of the original flywheel in the HEV was 600 N·m/rad. Considering the influence of various stiffness, the authors set the TS of the DMF to 60, 100, 150, and 200 N·m/rad.

With different kinds of damping of the DMF, the TV responses of the carrier and the differential under the excitation of the engine are demonstrated in Figs. 6.27 and 6.28.

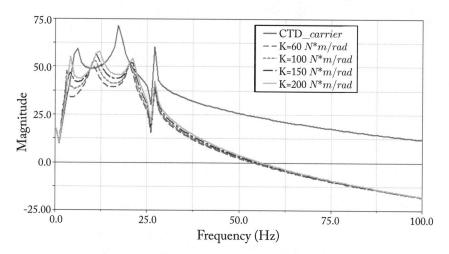

Figure 6.25: Frequency response of carrier with the different stiffness of flywheels.

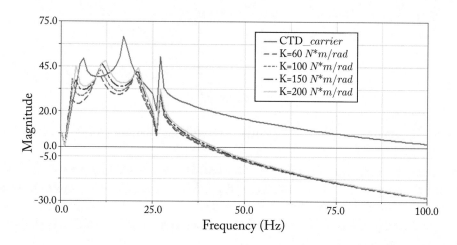

Figure 6.26: Frequency response of differential with different stiffness of flywheels.

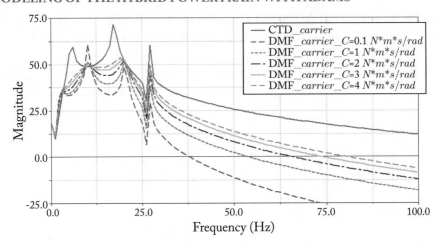

Figure 6.27: Frequency response of carrier with different damping of flywheels.

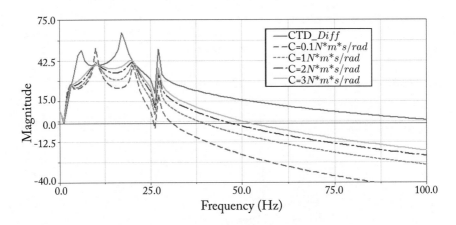

Figure 6.28: Frequency response of differential with different damping of flywheels.

The maximum response curve in Figs. 6.25 and 6.26 is the TV response of the DMF. We can see form the figures, the smaller the TS of the DMF, the smaller the amplitude-frequency response of the powertrain components. In particular, the response amplitude in the low frequency phase is greatly reduced. Consequently, when designing the TS of a DMF, the TS should be reduced as low as possible.

6.4.5   INFLUENCE OF THE DAMPING OF DMF

The damping energy consumes the vibration energy of the system, which can reduce the amplitude of the system resonance so that the system vibration converges and the amplitude decays. If the damping is too small, the TV of the system cannot be effectively attenuated. If the damping is too large, the TS of the system is increased and the torsional performance is reduced. Consequently, choosing a reasonable damping value is very important for the design of the DMF.

Second, the damping of DMF is optimized by a FV analysis. In the simulation, the damping value of the DMF is set to 0.1, 1, 2, 3, and 4 N·m·s/rad.

With different kinds of damping of the DMF, the TV responses of the carrier and the differential under the excitation of the engine are demonstrated in Figs. 6.27 and 6.28.

As mentioned, it is not effective to decrease the resonance amplitude when the damping is 0.1 N·m·s/rad. Besides, the damping increases the TVs of the powertrain components in the most ranges of frequency. Comparing the results for various damping values reveals, the TV response of the hybrid powertrain is fairly enhanced when the damping is 1 N·m·s/rad.

6.5   CONCLUSIONS

The aim of this section is to explore the TV performance for a power-split hybrid powertrain equipped with a DMF and provides a valuable reference for the dynamic design of hybrid vehicles. In the model study, a multi-body dynamic model of the hybrid powertrain is established and verified by mathematical methods for the prediction of the TV characteristics of this powertrain. The results illustrate that the hybrid powertrain with a DMF plays a critical role in decreasing the TV comparing with a powertrain equipped with the traditional flywheel. Further, the outcomes demonstrate that the amplitude of the TVs is fairly enhanced when the rotational inertias of the primary and the secondary flywheels are 0.21 kg·m^2 and 0.07 kg·m^2, respectively, and the TS and damping of the DMF are 60 N·m/rad and 1 N·m·s/rad, respectively.

References

[1] Chan, C. C. The state of the art of electric and hybrid vehicles. *Proc. of the IEEE*, 90.2 (2002): 247–275. DOI: 10.1109/5.989873. 1

[2] Fredriksson, Jonas, Henrik Weiefors, and Bo Egardt. Powertrain control for active damping of driveline oscillations. *Vehicle System Dynamics*, 37.5 (2002): 359–376. DOI: 10.1076/vesd.37.5.359.3527.

[3] Theodossiades, Stephanos, et al. Mode identification in impact-induced high-frequency vehicular driveline vibrations using an elasto-multi-body dynamics approach. *Proc. of the Institution of Mechanical Engineers, Part K: Journal of Multi-body Dynamics*, 218.2 (2004): 81–94. DOI: 10.1243/146441904323074549.

[4] Menday, M. T., Homer Rahnejat, and M. Ebrahimi. Clonk: An onomatopoeic response in torsional impact of automotive drivelines. *Proc. of the Institution of Mechanical Engineers, Part D: Journal of Automobile Engineering*, 213.4 (1999): 349–357. DOI: 10.1243/0954407991526919.

[5] Duoba, Michael, Henry Ng, and Robert Larsen. In-situ mapping and analysis of the Toyota Prius HEV engine, no. 2000–01-3096. SAE Technical Paper, 2000. DOI: 10.4271/2000-01-3096.

[6] Miller, John M. Hybrid electric vehicle propulsion system architectures of the e-CVT type. *IEEE Transactions on Power Electronics*, 21.3 (2006): 756–767. DOI: 10.1109/tpel.2006.872372.

[7] Ahn, Kukhyun, Sungtae Cho, and Suk Won Cha. Optimal operation of the power-split hybrid electric vehicle powertrain. *Proc. of the Institution of Mechanical Engineers, Part D: Journal of Automobile Engineering*, 222.5 (2008): 789–800. DOI: 10.1243/09544070jauto426.

[8] Canova, Marcello, Yann Guezennec, and Steve Yurkovich. On the control of engine start/stop dynamics in a hybrid electric vehicle. *Journal of Dynamic Systems, Measurement, and Control*, 131.6 (2009): 061005. DOI: 10.1115/1.4000066. 1

[9] Watts, G. R. A comparison of noise measures for assessing vehicle noisiness. *Journal of Sound and Vibration*, 180.3 (1995): 493–512. DOI: 10.1006/jsvi.1995.0092. 1

[10] Zhang, Xiaowu, Huei Peng, and Jing Sun. A near-optimal power management strategy for rapid component sizing of multimode power split hybrid vehicles. *IEEE Transactions on Control Systems Technology*, 23.2 (2015): 609–618. DOI: 10.1109/tcst.2014.2335060.

[11] Shiau, Ching-Shin Norman, et al. Optimal plug-in hybrid electric vehicle design and allocation for minimum life cycle cost, petroleum consumption, and greenhouse gas emissions. *Journal of Mechanical Design*, 132.9 (2010): 091013. DOI: 10.1115/1.4002194.

[12] Meng, Fei, et al. System modeling, coupling analysis, and experimental validation of a proportional pressure valve with pulsewidth modulation control. *IEEE/ASME Transactions on Mechatronics*, 21.3 (2016): 1742–1753. DOI: 10.1109/tmech.2015.2499270.

[13] Kuang, Ming L. An investigation of engine start-stop NVH in a power split powertrain hybrid electric vehicle, no. 2006–01-1500. SAE Technical Paper, 2006. DOI: 10.4271/2006-01-1500.

[14] Schulz, Marcus. Low-frequency torsional vibrations of a power split hybrid electric vehicle drive train. *Modal Analysis*, 11.6 (2005): 749–780. DOI: 10.1177/1077546305053661. 1

[15] Huang, Yanjun, et al. A review of power management strategies and component sizing methods for hybrid vehicles. *Renewable and Sustainable Energy Reviews*, 96 (2018): 132–144. DOI: 10.1016/j.rser.2018.07.020. 1

[16] Huang, Yanjun, et al. Model predictive control power management strategies for HEVs: A review. *Journal of Power Sources*, 341 (2017): 91–106. DOI: 10.1016/j.jpowsour.2016.11.106.

[17] Qin, Yechen, et al. A novel global sensitivity analysis on the observation accuracy of the coupled vehicle model. *Vehicle System Dynamics* (2018): 1–22. DOI: 10.1080/00423114.2018.1517219.

[18] Qin, Yechen, et al. Speed independent road classification strategy based on vehicle response: Theory and experimental validation. *Mechanical Systems and Signal Processing*, 117 (2019): 653–666. DOI: 10.1016/j.ymssp.2018.07.035.

[19] Zou, Changfu, et al. Electrothermal dynamics-conscious lithium-ion battery cell-level charging management via state-monitored predictive control. *Energy*, 141 (2017): 250–259. DOI: 10.1016/j.energy.2017.09.048.

[20] Xiaolin Tang, et al. Research on the energy control of a dual-motor hybrid vehicle during engine start-stop process. *Energy*, 166 (2019): 1181–1193. DOI: 10.1016/j.energy.2018.10.130. 2

[21] Zou, Changfu, et al. Nonlinear fractional-order estimator with guaranteed robustness and stability for Lithium-Ion batteries. *IEEE Transactions on Industrial Electronics*, 65.7 (2018): 5951–5961. DOI: 10.1109/tie.2017.2782691. 1

[22] Duoba, Michael, Henry Ng, and Robert Larsen. Characterization and comparison of two hybrid electric vehicles (HEVs)—Honda Insight and Toyota Prius, no. 2001-01-1335. SAE Technical Paper, 2001. DOI: 10.4271/2001-01-1335. 1

[23] Cho, Sungtae, Kukhyun Ahn, and Jang Moo Lee. Efficiency of the planetary gear hybrid powertrain. *Proc. of the Institution of Mechanical Engineers, Part D: Journal of Automobile Engineering*, 220.10 (2006): 1445–1454. DOI: 10.1243/09544070jauto176.

[24] Ahn, Kukhyun, et al. Performance analysis and parametric design of the dual-mode planetary gear hybrid powertrain. *Proc. of the Institution of Mechanical Engineers, Part D: Journal of Automobile Engineering*, 220.11 (2006): 1601–1614. DOI: 10.1243/09544070jauto334.

[25] Bellomo, Pietro, et al. Innovative vehicle powertrain systems engineering: beating the noisy offenders in vehicle transmissions, no. 2000-01-0033. SAE Technical Paper, 2000. DOI: 10.4271/2000-01-0033.

[26] Eisele, Georg, et al. Application of vehicle interior noise simulation (VINS) for NVH analysis of a passenger car, no. 2005-01-2514. SAE Technical Paper, 2005. DOI: 10.4271/2005-01-2514.

[27] Govindswamy, Kiran, Thomas Wellmann, and Georg Eisele. Aspects of NVH integration in hybrid vehicles. *SAE International Journal of Passenger Cars-Mechanical Systems*, 2.2009-01-2085 (2009): 1396-1405. DOI: 10.4271/2009-01-2085.

[28] Syed, Fazal U., Ming L. Kuang, and Hao Ying. Active damping wheel-torque control system to reduce driveline oscillations in a power-split hybrid electric vehicle. *IEEE Transactions on Vehicular Technology*, 58.9 (2009): 4769–4785. DOI: 10.1109/tvt.2009.2025953.

[29] Guo, Yichao and Robert G. Parker. Purely rotational model and vibration modes of compound planetary gears. *Mechanism and Machine Theory*, 45.3 (2010): 365–377. DOI: 10.1016/j.mechmachtheory.2009.09.001.

[30] Parker, Robert G. and Xionghua Wu. Vibration modes of planetary gears with unequally spaced planets and an elastic ring gear. *Journal of Sound and Vibration*, 329.11 (2010): 2265–2275. DOI: 10.1016/j.jsv.2009.12.023.

[31] Shin, Won, et al. 6 speed automatic transmission vibration magnitude prediction and whine noise improvement through transmission system modeling, no. 2011-01-1553. SAE Technical Paper, 2011. DOI: 10.4271/2011-01-1553. 1

[32] Kahraman, A. Natural modes of planetary gear trains. *Journal of Sound Vibration*, 173 (1994): 125–130. DOI: 10.1006/jsvi.1994.1222. 1

[33] Sun, Tao and HaiYan Hu. Nonlinear dynamics of a planetary gear system with multiple clearances. *Mechanism and Machine Theory*, 38.12 (2003): 1371–1390. DOI: 10.1016/s0094-114x(03)00093-4.

[34] Eisele, Georg, et al. Application of vehicle interior noise simulation (VINS) for NVH analysis of a passenger car, no. 2005–01-2514. SAE Technical Paper, 2005. DOI: 10.4271/2005-01-2514.

[35] Zhang, Lei, et al. Experimental impedance investigation of an ultracapacitor at different conditions for electric vehicle applications. *Journal of Power Sources*, 287 (2015): 129–138. DOI: 10.1016/j.jpowsour.2015.04.043.

[36] Zhang, Jianwu, et al. Multi-body dynamics and noise analysis for the torsional vibration of a power-split hybrid driveline. *Proc. of the Institution of Mechanical Engineers, Part K: Journal of Multi-body Dynamics*, 228.4 (2014): 366–379. DOI: 10.1177/1464419314540152. 1, 20, 26, 28, 34, 40, 70, 86

[37] Li, R., Wang, J. *Gear System Dynamics*, Science Press, 1997 (in Chinese). 1, 51, 52, 55, 57, 59, 84

[38] Wang, Yongliang, et al. Design and analysis of a multi-stage torsional stiffness dual mass flywheel based on vibration control. *Applied Acoustics*, 104 (2016): 172–181. DOI: 10.1016/j.apacoust.2015.11.004.

[39] Pfleghaar, Joachim and Boris Lohmann. The electrical dual mass flywheel-an efficient active damping system. *IFAC Proceedings Volumes*, 46.21 (2013): 483–488. DOI: 10.3182/20130904-4-jp-2042.00046.

[40] Walter, Andreas, et al. Anti-jerk and idle speed control with integrated sub-harmonic vibration compensation for vehicles with dual mass flywheels. *SAE International Journal of Fuels and Lubricants*, 1.1 (2009): 1267–1276. DOI: 10.4271/2008-01-1737.

[41] Tang, Xiaolin, et al. Novel mathematical modelling methods of comprehensive mesh stiffness for spur and helical gears. *Applied Mathematical Modelling*, 64 (2018): 524–540. DOI: 10.1016/j.apm.2018.08.003.

[42] Tang, Xiaolin, et al. A novel simplified model for torsional vibration analysis of a series-parallel hybrid electric vehicle. *Mechanical Systems and Signal Processing*, 85 (2017): 329–338. DOI: 10.1016/j.ymssp.2016.08.020. 1

[43] Tang, Xiaolin, et al. Novel torsional vibration modeling and assessment of a power-split hybrid electric vehicle equipped with a dual-mass flywheel. *IEEE Transactions on Vehicular Technology*, 67.3 (2018): 1990–2000. DOI: 10.1109/tvt.2017.2769084. 9, 88, 105

[44] Qin, Yechen, et al. A novel nonlinear road profile classification approach for controllable suspension system: Simulation and experimental validation. *Mechanical Systems and Signal Processing* (2018). DOI: 10.1016/j.ymssp.2018.07.015.

[45] Zhang, Xi, Xiaolin Tang, and Wei Yang. Analysis of transmission error and load distribution of a hoist two-stage planetary gear system. *Proc. of the Institution of Mechanical Engineers, Part K: Journal of Multi-body Dynamics* (2018): 1464419318770886. DOI: 10.1177/1464419318770886.

[46] Yang, Wei and Xiaolin Tang. Modelling and modal analysis of a hoist equipped with two-stage planetary gear transmission system. *Proc. of the Institution of Mechanical Engineers, Part K: Journal of Multi-body Dynamics*, 231.4 (2017): 739–749. DOI: 10.1177/1464419316684067.

[47] Yang, Wei, Xiaolin Tang, and Xiaoan Chen. Nonlinear modelling and transient dynamics analysis of a hoist equipped with a two-stage planetary gear transmission system. *Journal of Vibroengineering*, 17.6 (2015).

[48] Huang, Yanjun, et al. A comparative study of the energy-saving controllers for automotive air-conditioning/refrigeration systems. *Journal of Dynamic Systems, Measurement, and Control*, 139.1 (2017): 014504. DOI: 10.1115/1.4034505. 9

[49] Tang, Xiaolin, et al. Torsional vibration and acoustic noise analysis of a compound planetary power-split hybrid electric vehicle. *Proc. of the Institution of Mechanical Engineers, Part D: Journal of Automobile Engineering*, 228.1 (2014): 94–103. DOI: 10.1177/0954407013508276. 5, 19, 25, 29, 64, 69

[50] Tang, Xiaolin, et al. Study on the torsional vibration of a hybrid electric vehicle powertrain with compound planetary power-split electronic continuous variable transmission. *Proc. of the Institution of Mechanical Engineers, Part C: Journal of Mechanical Engineering Science*, 228.17 (2014): 3107–3115. DOI: 10.1177/0954406214526162. 6

[51] Huang, Yanjun, et al. A supervisory energy-saving controller for a novel anti-idling system of service vehicles. *IEEE/ASME Transactions on Mechatronics*, 22.2 (2017): 1037–1046. DOI: 10.1109/tmech.2016.2631897.

[52] Huang, Yanjun, et al. An energy-saving set-point optimizer with a sliding mode controller for automotive air-conditioning/refrigeration systems. *Applied Energy*, 188 (2017): 576–585. DOI: 10.1016/j.apenergy.2016.12.033.

[53] Qin, Yechen, et al. Vibration mitigation for in-wheel switched reluctance motor driven electric vehicle with dynamic vibration absorbing structures. *Journal of Sound and Vibration*, 419 (2018): 249–267. DOI: 10.1016/j.jsv.2018.01.010.

[54] Qin, Yechen, et al. Comprehensive analysis for influence of controllable damper time delay on semi-active suspension control strategies. *Journal of Vibration and Acoustics*, 139.3 (2017): 031006. DOI: 10.1115/1.4035700.

[55] Qin, Yechen, et al. Road excitation classification for semi-active suspension system based on system response. *Journal of Vibration and Control*, 24.13 (2018): 2732–2748. DOI: 10.1177/1077546317693432. 19

[56] Hu, Chuan, et al. Differential steering based yaw stabilization using ISMC for independently actuated electric vehicles. *IEEE Transactions on Intelligent Transportation Systems*, 19.2 (2018): 627–638. DOI: 10.1109/tits.2017.2750063.

[57] Hwang, Sheng-Jiaw, Joseph L. Stout, and Ching-Chung Ling. Modeling and analysis of powertrain torsional response, no. 980276. SAE Technical Paper, 1998. DOI: 10.4271/980276.

[58] Wang, Jianjun, Runfang Li, and Xianghe Peng. Survey of nonlinear vibration of gear transmission systems. *Applied Mechanics Reviews*, 56.3 (2003): 309–329. DOI: 10.1115/1.1555660. 53

[59] Yu, Hai-Sheng, Jian-Wu Zhang, and Tong Zhang. Control strategy design and experimental research on a four-shaft electronic continuously variable transmission hybrid electric vehicle. *Proc. of the Institution of Mechanical Engineers, Part D: Journal of Automobile Engineering*, 226.12 (2012): 1594–1612. DOI: 10.1177/0954407012450117.

[60] Eisele, Georg, et al. Acoustics of hybrid vehicles, no. 2010–01-1402. SAE Technical Paper, 2010. DOI: 10.4271/2010-01-1402.

[61] Li, Liang, et al. Correctional DP-based energy management strategy of plug-in hybrid electric bus for city-bus route. *IEEE Transactions on Vehicular Technology*, 64.7 (2015): 2792–2803. DOI: 10.1109/tvt.2014.2352357.

[62] Yang, Yalian, et al. Comparison of power-split and parallel hybrid powertrain architectures with a single electric machine: dynamic programming approach. *Applied Energy*, 168 (2016): 683–690. DOI: 10.1016/j.apenergy.2016.02.023.

[63] Tang, Xiaolin, et al. A novel two degree-of-freedom dynamic model of a full hybrid vehicle. *International Journal of Electric and Hybrid Vehicles*, 9.1 (2017): 67–77. DOI: 10.1504/ijehv.2017.10003710.

[64] Tang, Xiaolin, et al. Torsional vibration characteristics of a power–split hybrid system. *International Journal of Electric and Hybrid Vehicles*, 5.2 (2013): 108–122. DOI: 10.1504/ijehv.2013.056294. 53

[65] Yang, Wei and Xiaolin Tang. Numerical analysis for heat transfer laws of a wet multi-disk clutch during transient contact. *International Journal of Nonlinear Sciences and Numerical Simulation*, 18.7–8 (2017): 599–613. DOI: 10.1515/ijnsns-2017-0081. 63

[66] Huang, Yanjun, et al. An adaptive model predictive controller for a novel battery-powered anti-idling system of service vehicles. *Energy*, 127 (2017): 318–327. DOI: 10.1016/j.energy.2017.03.119.

[67] Huang, Yanjun, et al. Modelling and optimal energy-saving control of automotive air-conditioning and refrigeration systems. *Proc. of the Institution of Mechanical Engineers, Part D: Journal of Automobile Engineering*, 231.3 (2017): 291–309. DOI: 10.1177/0954407016636978.

[68] Huang, Yanjun, et al. Optimal energy-efficient predictive controllers in automotive air-conditioning/refrigeration systems. *Applied Energy*, 184 (2016): 605–618. DOI: 10.1016/j.apenergy.2016.09.086.

[69] Huang, Yanjun, Amir Khajepour, and Hong Wang. A predictive power management controller for service vehicle anti-idling systems without a priori information. *Applied Energy*, 182 (2016): 548–557. DOI: 10.1016/j.apenergy.2016.08.143.

[70] Qin, Yechen, et al. Road excitation classification for semi-active suspension system with deep neural networks. *Journal of Intelligent and Fuzzy Systems*, 33.3 (2017): 1907–1918. DOI: 10.3233/jifs-161860.

[71] Wang, Zhen-Feng, et al. Influence of road excitation and steering wheel input on vehicle system dynamic responses. *Applied Sciences*, 7.6 (2017): 570. DOI: 10.3390/app7060570.

[72] Kahraman, Ahmet. Free torsional vibration characteristics of compound planetary gear sets. *Mechanism and Machine Theory*, 36.8 (2001): 953–971. DOI: 10.1016/s0094-114x(01)00033-7. 63, 65

[73] Qin, Yechen, et al. Adaptive hybrid control of vehicle semiactive suspension based on road profile estimation. *Shock and Vibration* (2015). DOI: 10.1155/2015/636739. 63

[74] Miller, John M. and Michael Everett. An assessment of ultra-capacitors as the power cache in Toyota THS-II, GM-Allison AHS-2 and Ford FHS hybrid propulsion systems. *Applied Power Electronics Conference and Exposition (APEC). 20th Annual IEEE*, vol. 1, 2005. DOI: 10.1109/apec.2005.1452980.

[75] Gelb, George H., et al. An electromechanical transmission for hybrid vehicle power trains-design and dynamometer testing, no. 710235. SAE Technical Paper, 1971. DOI: 10.4271/710235.

[76] Kubur, M., et al. Dynamic analysis of a multi-shaft helical gear transmission by finite elements: model and experiment. *Journal of Vibration and Acoustics*, 126.3 (2004): 398–406. DOI: 10.1115/1.1760561.

[77] Gnanakumarr, M., et al. Impact-induced vibration in vehicular driveline systems: theoretical and experimental investigations. *Proc. of the Institution of Mechanical Engineers, Part K: Journal of Multi-body Dynamics*, 219.1 (2005): 1–12. DOI: 10.1243/1464419053577626.

[78] Miller, John M., Patrick J. McCleer, and Michael Everett. Comparative assessment of ultra-capacitors and advanced battery energy storage systems in PowerSplit electronic-CVT vehicle powertrains. *Electric Machines and Drives, IEEE International Conference on*, 2005. DOI: 10.1109/iemdc.2005.195921.

[79] Kim, Hyung Min, et al. Target cascading in optimal system design. *Journal of Mechanical Design*, 125.3 (2003): 474–480. DOI: 10.1115/1.1582501.

[80] Rook, T. E. and R. Singh. Dynamic analysis of a reverse-idler gear pair with concurrent clearances. *Journal of Sound and Vibration*, 182.2 (1995): 303–322. DOI: 10.1006/jsvi.1994.0198. 63

[81] Hong, Qing-quan and Ying Cheng. Dynamic simulation of multistage gear train system in Adams. *Journal of Beijing Institute of Technology*, 6 (2003): 007 (in Chinese). 83, 84

[82] Theodossiades, Stephanos, et al. Effect of a dual-mass flywheel on the impact-induced noise in vehicular powertrain systems. *Proc. of the Institution of Mechanical Engineers, Part D: Journal of Automobile Engineering*, 220.6 (2006): 747–761. DOI: 10.1243/09544070jauto55. 100, 101

[83] Walker, Paul D. and Nong Zhang. Modelling of dual clutch transmission equipped powertrains for shift transient simulations. *Mechanism and Machine Theory*, 60 (2013): 47–59. DOI: 10.1016/j.mechmachtheory.2012.09.007.

[84] Song, Li Quan, et al. Design and analysis of a dual mass flywheel with continuously variable stiffness based on compensation principle. *Mechanism and Machine Theory*, 79 (2014): 124–140. DOI: 10.1016/j.mechmachtheory.2014.04.004.

[85] Read, M. G., R. A. Smith, and K. R. Pullen. Optimisation of flywheel energy storage systems with geared transmission for hybrid vehicles. *Mechanism and Machine Theory*, 87 (2015): 191–209. DOI: 10.1016/j.mechmachtheory.2014.11.001.

[86] Zeng, Li Ping, Li Quan Song, and Jian Dong Zhou. Design and elastic contact analysis of a friction bearing with shape constraint for promoting the torque characteristics of a dual mass flywheel. *Mechanism and Machine Theory*, 92 (2015): 356–374. DOI: 10.1016/j.mechmachtheory.2015.06.002.

[87] Kang, T. S., S. K. Kauh, and K. P. Ha. Development of the displacement measuring system for a dual mass flywheel in a vehicle. *Proc. of the Institution of Mechanical Engineers, Part D: Journal of Automobile Engineering*, 223.10 (2009): 1273–1281. DOI: 10.1243/09544070jauto1066.

[88] Littlefair, G. P. Arc spring sliding friction within the dual mass flywheel, M.Sc. Thesis, Loughborough University, Loughborough, UK, 2004.

[89] Hong, S. Dual mass flywheel using air damping, United States Patent, 2003/0233907. 104, 106

[90] Feldhaus, R. Two-mass flywheel, United States Patent, 5307710. 100

[91] LV, C. F. The characteristics study on torsional vibration of dual mass flywheel and its simulation analysis, Shanghai Jiao Tong University, 2008 (in Chinese). 104, 106

Authors' Biographies

XIAOLIN TANG

Xiaolin Tang received a B.S. in mechanics engineering and an M.S. in vehicle engineering from Chongqing University, China, in 2006 and 2009, respectively. He received a Ph.D. in mechanical engineering from Shanghai Jiao Tong University, China, in 2015. He is currently an Associate Professor at the State Key Laboratory of Mechanical Transmissions and at the Department of Automotive Engineering, Chongqing University, Chongqing, China. He is also a committeeman of Technical Committee on Vehicle Control and Intelligence of Chinese Association of Automation (CAA). He has led and has been involved in more than 10 research projects, such as National Natural Science Foundation of China, and has published more than 20 papers. His research focuses on Hybrid Electric Vehicles (HEVs), vehicle dynamics, noise and vibration, and transmission control.

YANJUN HUANG

Yanjun Huang is a Postdoctoral Fellow at the Department of Mechanical and Mechatronics Engineering at University of Waterloo, where he received his Ph.D. in 2016. His research interest is mainly on the vehicle holistic control in terms of safety, energy-saving, and intelligence, including vehicle dynamics and control, HEV/EV optimization and control, motion planning and control of connected and autonomous vehicles, and human-machine cooperative driving. He has published several books and over 50 papers in journals and conferences. He currently serves as an associate editor and editorial board member of *IET Intelligent Transport System*, *SAE International Journal of Commercial vehicles*, *International Journal of Vehicle Information and Communications*, *Automotive Innovation*, *AIME*, etc.

HONG WANG

Hong Wang is currently a research associate of Mechanical and Mechatronics Engineering with the University of Waterloo. She received her Ph.D. from the Beijing Institute of Technology in China in 2015. Her research focuses on the component sizing, modeling of hybrid powertrains, and energy management control strategies design for Hybrid electric vehicles; intelligent control theory and application; and autonomous vehicles.

YECHEN QIN

Yechen Qin is currently a Postdoctoral Fellow of mechanical engineering with the Beijing Institute of Technology, where he received his B. Eng and Ph.D. in 2010 and 2016. From 2013–2014, he studied at Texas A&M University as a visiting Ph.D. student. From 2017–2018, he studied at the University of Waterloo as a visiting scholar. His research interests include vehicle dynamics control, road estimation, and in-wheel motor vibration control.

Printed in the United States
by Baker & Taylor Publisher Services